Lecture Notes in Electrical Engineering

Volume 299

For further volumes:
http://www.springer.com/series/7818

About this Series

"Lecture Notes in Electrical Engineering (LNEE)" is a book series which reports the latest research and developments in Electrical Engineering, namely:

- Communication, Networks, and Information Theory
- Computer Engineering
- Signal, Image, Speech and Information Processing
- Circuits and Systems
- Bioengineering

LNEE publishes authored monographs and contributed volumes which present cutting edge research information as well as new perspectives on classical fields, while maintaining Springer's high standards of academic excellence. Also considered for publication are lecture materials, proceedings, and other related materials of exceptionally high quality and interest. The subject matter should be original and timely, reporting the latest research and developments in all areas of electrical engineering.

The audience for the books in LNEE consists of advanced level students, researchers, and industry professionals working at the forefront of their fields. Much like Springer's other Lecture Notes series, LNEE will be distributed through Springer's print and electronic publishing channels.

Radhakrishna Maringanti · Murlidhar Tiwari
Anish Arora
Editors

Proceedings of Ninth International Conference on Wireless Communication and Sensor Networks

WCSN 2013

Springer

Editors
Radhakrishna Maringanti
Electronics and Communication Engineering
Indian Institute of Information Technology
Allahabad
Uttar Pradesh
India

Anish Arora
Computer Science and Engineering and Institute of Sensing System
Ohio State University
Columbus, OH
USA

Murlidhar Tiwari
Administration Block
Indian Institute of Information Technology
Allahabad
Uttar Pradesh
India

ISSN 1876-1100 ISSN 1876-1119 (electronic)
ISBN 978-81-322-1822-7 ISBN 978-81-322-1823-4 (eBook)
DOI 10.1007/978-81-322-1823-4
Springer New Delhi Heidelberg New York Dordrecht London

Library of Congress Control Number: 2014934689

Printed on acid-free paper

Springer is part of Springer Science+Business Media (www.springer.com)

Preface

Dear Reader,

Welcome to the Proceedings of the 9th International Conference on Wireless Communication and Sensor Networks (WCSN 2013) organized by the Indian Institute of Information Technology, Allahabad.

A number of developments in the area of Wireless Communication Technologies, Embedded Processors, Semantic Web, Smart Surroundings, Network Devices, and Sensor Networks have taken place over the past 9 years in the broad area of Wireless Communication and Sensor Networks, and these make it possible to take up real life applications. These technologies have reached a level of maturity that would allow the creation of large, reliable, and pervasive services for the benefit of the society. Thus, Wireless Communication and Sensor Networks is of great interest to the research community due to its promised impact on the society. Wireless Sensor and Actor Networks (WSAN) not only promise to influence the world by their pervasive presence in the remotest locations, but also promise distributed monitoring and control.

The power-aware designs of network components, availability of adequate wireless channel bandwidths, availability of reliable sensor network components, and sensor network technologies enable us to build large and sustainable application systems that would benefit the society. Application areas such as health, agriculture, environment monitoring, industrial monitoring, structural monitoring, real-time tracking, and many other similar areas are looking for solutions from pervasive computing and wireless sensor networks.

If the sensor networks are to be used in real-life applications beyond the "concept proving" and "laboratory experimentation" stages, further investigations are needed on a number of challenging development issues. There are also challenges in developing intelligent, distributed, and collaborative processing, deployment of multimodal networks of significant size that would sense and act in wide areas in an unattended and a reliable manner.

The Ninth International Wireless Communication and Sensor Networks (WCSN 2013) Conference is organized to facilitate the information exchange with regard to the development of technologies, applications, and experiences with focus on large deployable applications.

A total of about 191 papers were received. Keeping in view the relevance, quality, and plagiarism issues, with the help of at least 2 reviewers per paper a total of 27 papers were accepted for presenting in the conference and for publishing in proceedings. Out of these papers 5 papers were withdrawn and the remaining 22 papers were organized into the following sections, in accordance with the conference themes:

Track 1: Wireless Communications
Track 2: Devices, Tools, and Techniques for WSN and other wireless networks
Track 3: Wireless Sensor Networks
Track 4: Sustainable Pervasive WSN Applications

This volume would be useful for researchers working in this fascinating and fast growing research area.

Organizing Committee: WCSN-2013

Conference Chairman: Murlidhar Tiwari (Director, IIIT-A)
Conference Co-chairman: Radhakrishna Maringanti, IIIT-A
Organization Co-chair: Shekhar Verma, IIIT-A
Publicity Co-chairs: Manish Kumar, IIIT-A and Ashutosh Kumar, IIIT-A
Fund Raising Co-chair: Rajat Singh, IIIT-A
Exhibition Co-chair: Shirshu Varma, IIIT-A

Technical Advisory Committee: WCSN-2013

Ajit Chaturvedi (IIT, Kanpur, India)
Anish Arora (Ohio State University, USA)
Anurag Kumar (IISC, Bangalore, India)
Carla P. Gomes (Cornell University, USA)
Dharma P. Agrawal (University of Cincinnati, USA)
George Karagiannidis (Aristotle University of Thessaloniki, Greece)
Hari M. Gupta (IIT, Delhi, India)
Erol Gelenbe (Imperial College, London, UK)
Julie McCann (Imperial College, London, UK)
Kim Kiseon (Gwangju Institute of Science and Technology, Korea)
Radhakrishna Maringanti (IIIT, Allahabad, India)
Man-Gon Park (P K National University, Korea)
Mani Srivastava (UCLA, USA)
Pascal Lorenz (University of Haute Alsace, France)
P. Vijay Kumar (IISC, Bangalore, India)
Prem Kumar Kalra (Dayalbagh Educational Institute, Agra, India)
R. V. Raja Kumar (RGUKT, Andhra Pradesh, India)
Rajat Moona (CDAC, Pune, India)
Sajal Das (University Texas at Arlington, USA)
Sanjay Madria (Missouri University of Science and Technology, USA)
Shekhar Verma (IIIT, Allahabad, India)
Shankar Lall Maskara (Jaypee Institute of Information Technology, India)

Shirshu Varma (IIIT, Allahabad, India)
Torsten Braun (University of Bern, Switzerland)
Xiaochun Cheng (Middlesex University, UK)
Xun Yi (Victoria University, Australia)
Yatindra Nath Singh (IIT, Kanpur, India)

Reviewers: WCSN-2013

A. R. Harish (IIT, Kanpur, India)
Adrish Banerjee (IIT, Kanpur, India)
Ajit Chaturvedi (IIT, Kanpur, India)
Anish Arora (Ohio State University, USA)
Anurag Kumar (IISc, Bangalore, India)
Carla P. Gomes (Cornell University, USA)
Dharma P. Agrawal (University of Cincinnati, USA)
David Grace (University of York, UK)
Edith Ngai (Uppsala University, Sweden)
Erol Gelenbe (Imperial College, London, UK)
George Karagiannidis (Aristotle University of Thessaloniki, Greece)
Janet Light (UNB, New Brunswick, Canada)
Julie McCann (Imperial College, London, UK)
Radhakrishna Maringanti (IIIT, Allahabad, India)
Madhur Deo Upadhyay (Shiv Nadar University, Gr. Noida, India)
Manoj Singh Parihar (IIITDM, Jabalpur, India)
Mani Srivastava (UCLA, USA)
Niki Trigoni (Oxford University, UK)
Neetesh Purohit (IIIT, Allahabad, India)
P. Vijay Kumar (IISC Bangalore, India)
Pascal Lorenz (University of Haute Alsace, France)
Partha Sarthi Mandal (IIT, Guwahati, India)
Peter Jun/Jung Hyun Jun (Singapore University of Technology and Design, Singapore)
Pooja Jain (Jaypee University of Information Technology, Waknaghat, India)
Poonam Yadav (Imperial College, London, UK)
Prem Kumar Kalra (Dayalbagh Educational Institute, Agra, India)
R. V. Rajakumar (RGUKT, Andhra Pradesh, India)
Rajat Moona (CDAC, Pune, India)
Sajal Das (University Texas at Arlington, USA)
Sanjay Madria (Missouri University of Science and Technology, USA)
Shankar Lall Maskara (Jaypee Institute of Information Technology, Noida, India)

Shekhar Verma (IIIT, Allahabad, India)
Shirshu Varma (IIIT, Allahabad, India)
Torsten Braun (University of Bern, Switzerland)
U. S. Tiwari (IIIT, Allahabad, India)
Xun Yi (Victoria University, Australia)
Yatindra Nath Singh (IIT, Kanpur, India)

Acknowledgments

We gratefully acknowledge the contribution of Indian Institute of Information Technology-Allahabad (IIIT-A) for the success of the conference. It has contributed enormously over the past 9 years by nurturing the conference from the early stages till it grew to the current position of one of the important conferences in the area of Wireless Sensor Networks. We sincerely thank the entire family of IIIT-Allahabad including faculty, research scholars, staff, and students for their assistance and hard work in organizing the conference and its proceedings year after year. We gratefully acknowledge the continuing support provided by the Department of Science and Technology (DST) and Department of Electronics and information Technology (DeitY), in organizing the conference. We also thank the Patent Referral Centre, Indian Institute of Information Technology, Allahabad, setup by the ministry of Communication and Information Technology, Government of India for conducting plagiarism check on all the technical papers submitted.

We also express our deep sense of gratitude to all the invited speakers, workshop presenters, and authors without whose support, this conference would not have been in this position.

We profusely thank the members of the technical advisory committee, the reviewers, and the members of the organizing committee for their support to the conference.

We also acknowledge the support provided by Publishers Springer to the WCSN-2013 by bringing out proceedings of WCSN-2013.

Radhakrishna Maringanti
Murlidhar Tiwari
Anish Arora

Contents

Chapter 1
Optimum Detection Probability with Partially Controlled Random Deployment of Wireless Sensors with Mobile Base Stations

Junghyun Jun and Dharma P. Agrawal

Abstract In this paper we analyze the problem of covering widely expanded field with wireless sensors where many of the known deployment and data aggregation methods become impractical. We deploy the wireless sensors in a partially controlled manner such that they are randomly placed on the lines of grid and mobile base stations like UAVs could be used to collect the data from the wireless sensors. Our objective is to maximize the detection probability of an event without overly deploying the sensors on the field. We have defined the detection probability to be the product of probability of an event been sensed and that data being collected by an UAV. Under this model, we analytically obtain a relationship between the grid spacing and a number of available UAVs which can maximize detection probability when two collaborative and independent strategies for UAVs and obtain some useful relationship in guiding design specification.

Keywords Wireless sensor network • Mobile base stations • Large-scale network • Optimum controlled random deployment

1.1 Introduction

Coverage in wireless sensor networks (WSNs) is one of the key parameters since detecting an event in the field of interest is one of its main objectives. The coverage can be defined as a probability of an event being detected by at least one sensor within the deployed area. Even though controlled deployments need fewer sensors

---

J. Jun (✉) · D. P. Agrawal
OBR Center of Distributed and Mobile Computing, Department of Computer Science, University of Cincinnati, 45221-0030 Cincinnati, OH, USA
e-mail: junjn@mail.uc.edu

D. P. Agrawal
e-mail: dpa@cs.uc.edu

R. Maringanti et al. (eds.), *Proceedings of Ninth International Conference on Wireless Communication and Sensor Networks*, Lecture Notes in Electrical Engineering 299, DOI: 10.1007/978-81-322-1823-4_1, © Springer India 2014

to provide full coverage (see [1–3]), sensor deployment following regular pattern quickly becomes very impractical when it comes to a wider field, especially for irregular terrains like national parks. Therefore, a random deployment is used as an alternative method. Dropping the sensors from the arial transportation is a good example of the random deployment. A uniform random deployment is often used as a good model for randomly deployed sensors in the field [4–5]. Let us assume that the field can be represented by a square of length L and events are approximately in circular shape, such as forest fire, chemical spreading, or sound of endangered species. In this model, the coverage can be represented by the probability $P_{\text{random}} = 1 - \left(1 - \frac{\pi d^2}{L^2}\right)^n$, where d is the event radius, n is the number of sensors deployed in the field. Here, event radius is equivalent to sensing radius but defining it as an event radius coincides with our geographical problem formulation which becomes clear later on. Let $k = L/d$ which represents the scale of the field corresponding to the event radius. Then, in order to get the coverage of 0.9, around 1 million sensors must be deployed when $k = 1{,}000$. In words, if event radius is 10 m and field size is 10 by 10 km, we must randomly deploy 1 million sensors in order to get 90 % of coverage. Deploying 1 million sensors in a uniform manner is impractical if field size grows more than few hundred kilometers. Even if one drops them from the air it will not look as close to a uniform distribution. However, most of the coverage-related works focused on achieving the full coverage either by random or regular deployments and maintaining such coverage as long as possible or as good as possible [6–7]. But they do not talk about the alternative methods if it is not possible to achieve it from the beginning.

The lifetime of WSN is another problem for widely expanded field since redeploying or replacing dead sensors are very difficult tasks. The idea of multiple mobile units have been used to cover a large area by utilizing them as mobile relays [8, 9] for connecting the disconnected clusters or as data mules [10] that deliver data from sensors to a base station. This idea does improve the lifetime of the networks but does not solve the problem of covering all the areas since no sensors are deployed in between the clusters or in between sensors to base station. The coverage using mobile sensors network without any static sensors were analyzed by Bisnik [11]. His conclusion was that mobility improves the coverage. However, it requires continuously moving mobile sensors but energy depletion due to mobility was not taken into consideration. These motivated us to look at more practical method of deploying and managing the WSN for monitoring an extremely large area.

We soon realize that it requires a random deployment with some regularity for a widely expanded field. So, instead of a uniform random deployment, the wireless sensors are deployed as a $m \times m$ grid, as shown in Fig. 1.1, where each line represents a set of sensors. This can reduce the total number of sensors used and it is more practical since it just requires a single fly over and drops the sensors along the line. It is not necessary for sensors to be evenly spaced as along as there are enough sensors so that their average interspacing between adjacent pair of sensors are less than the event radius d. Leoncini is the first one to look at this type of partially controlled deployment of sensors for widely expanded field [12]. In their paper, sensors are deployed at designated square in the grid which follows

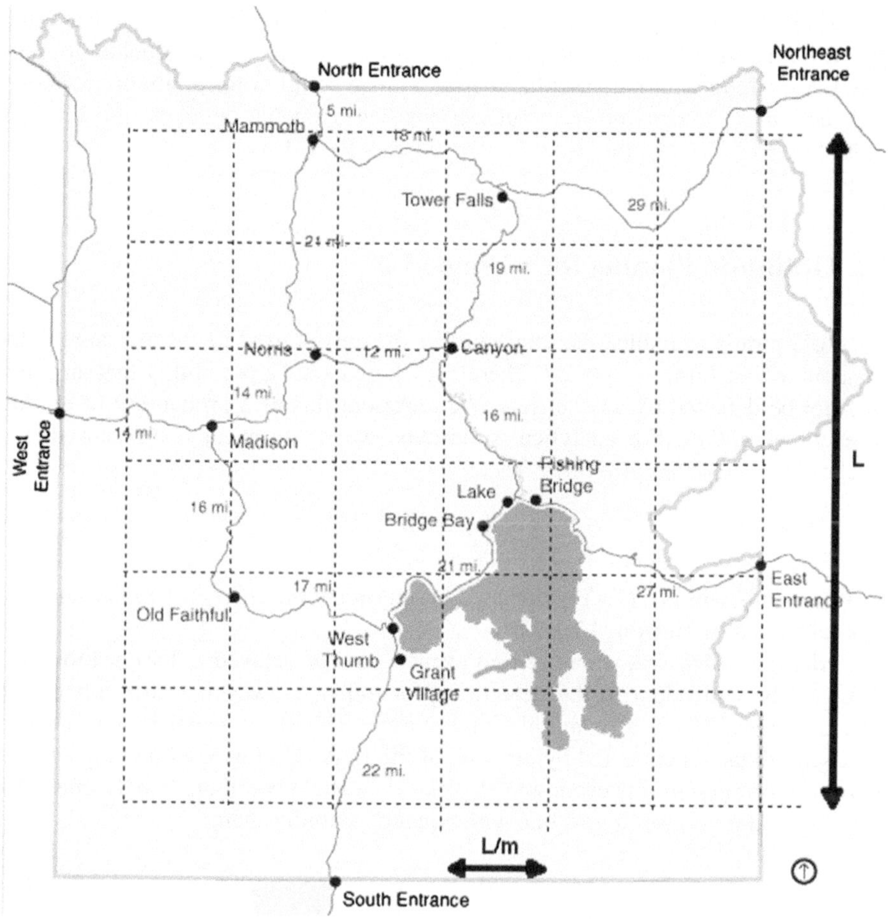

Fig. 1.1 The wireless sensors layout on Yellowstone National Park

a normal distribution whereas we deploy our sensors on the grid lines. However, the reduction of total number of sensors in this grid deployment comes at a cost of reducing the robustness of WSN since there are not too many alternative paths to a sink node (or base station) as failure of a small set of sensors can cause a network partitioning. In order to compensate for this problem, we introduce Unmanned Arial Vehicles (UAVs) as mobile base station. The idea is that UAVs will collect the data while they are traveling along a line which is selected randomly from the grid. The reason for using UAV is for its convenience since it would be difficult for ground vehicles to travel in a straight line following the grid. However, it is not necessary to use the UAVs as a mobile base station and choice of a mobile base station has no effect on our analysis. There is no data relaying in this method. Sensors transmit their data directly to flying UAVs when they are in its vicinity. In this paper, we determine the optimum grid spacing which can maximize the probability of an event getting detected by a given number of available UAVs.

The rest of this paper is organized as follows: The problem is formulated and solved for a single UAV in Sect. 1.2. In Sect. 1.3, we expand the problem to multiple UAVs along with numerical and approximate solution for optimum planning. The section is divided into two subsections based on the collaboration level of multiple UAVs. Finally, the paper is concluded in Sect. 1.4.

1.2 Optimum Planing for a Single UAV

Since our aim is to reduce the total number of sensors used to cover a large field, we assume that $L/(m-1) \geq 2d$. Therefore, there exists a possibility that an event may not be detected by any sensors. We represent this as a probability of getting an event sensed P_s. For analytical simplicity, we let an event being sensed if it touches any one of the grid lines. Then,

$$P_s = 1 - \left(1 - \frac{(m-1)d}{L}\right)^2. \tag{1.1}$$

This probability in Eq. (1.1) is obtained based on geometrical probability and it is first calculated by Buffon [13].

Under normal WSNs where sensed event would be relayed to base station, this P_s would be equivalent to the detection probability. However, in our model, we have employed UAVs as a mobile base station. Therefore, event is not detected until sensed data is collected by any one of the UAVs. Let us use "pick up" as our ways of representing this event and $P(\text{pickup}) = P_p$. When there is only one UAV available and it chooses a grid line with equal probability then,

$$P_p = \frac{1}{2m}. \tag{1.2}$$

Here, we define that

$$\begin{aligned} P_d &= P\,(\text{an event is detected}) \\ &= P\,(\text{an event is sensed and pick up by an UAV}). \end{aligned} \tag{1.3}$$

We assume that an event sensed by a grid line and picked up by a UAV are independent events and it is true if ones does not know the distribution of event locations. Therefore,

$$P_d = P_s P_p = \left(1 - \left(1 - \frac{(m-1)d}{L}\right)^2\right)\frac{1}{2m}. \tag{1.4}$$

Note an interesting paradox in Eq. (1.4) that increasing m helps P_s but reduces P_p and vice versa. This suggests that there must exist an optimum grid spacing which maximizes the detection probability. The optimum size m^* of grid can be solved easily by differentiating Eq. (1.4).

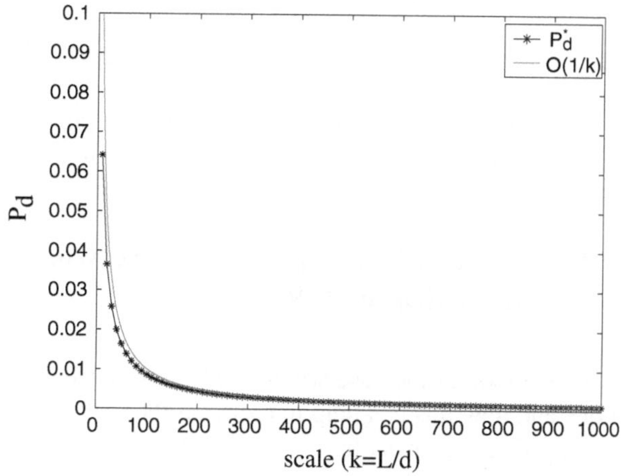

Fig. 1.2 The decreasing of maximum detection probability with a single UAV when scale representing the relative size of network grows. Graph also shows that the probability is decreasing approximately in $O(1/k)$

$$
\begin{aligned}
m^* &= \left\lfloor \sqrt{\frac{2L+d}{d}} \right\rfloor \\
&= \left\lfloor \sqrt{\frac{2L}{d}+1} \right\rfloor \\
&\approx \left\lfloor \sqrt{2k} \right\rfloor .
\end{aligned} \tag{1.5}
$$

The floor sign is placed over the optimum $m*$ since our aim is to reduce the total number of sensors and m can only be integer. However, ceiling could be another possible solution if the flooring reduces the detection probability significantly as compared to ceiling it. Note from the optimum size of m in Eq. (1.5) that it grows approximately square root of two times of k which represents the scale of field corresponding to event radius as defined earlier.

By substituting Eq. (1.5) without floor sign back into Eq. (1.4), we obtain maximum detection probability P_d* for a single UAV case.

$$
\begin{aligned}
P_d^* &= \frac{d}{L} + \frac{d^2}{L^2}\left(1 - \sqrt{\frac{2L+d}{d}}\right) \\
&= O\left(\frac{1}{k}\right).
\end{aligned} \tag{1.6}
$$

The maximum detection probability is numerically computed and plotted as shown in Fig. 1.2.

Suppose, a UAV flies over the $m* \times m*$ grid network at a fixed period of T. Then, probability that an event is detected before time s given an event has occur at time zero is geometrically distributed as,

$$P(t < s|s \approx nT) = \sum_{i=1}^{n} \left(1 - P_d^*\right)^{i-1} P_d^*. \tag{1.7}$$

With this, one can easily determine the suitable T and m which can satisfy their coverage demands under the constraint like a total number of sensors available.

1.3 Optimum Grid Spacing for Network with Multiple UAVs

Now we look at the case when n UAVs are available to be used as mobile base stations and how this can help in reducing the total number of sensors to be used, while maximizing the detection probability. We have looked at two different cases of multiple UAVs: n independent UAVs and n collaborative UAVs. The collaborative UAVs will not fly over the same grid line at the same time. Whereas, there is a good chance of selecting the same grid line in case of noncollaborative independent UAVs. It is obvious that a noncollaborative case will not perform as good as a collaborative case. However, in practice this situation can arise if each UAV is owned by different organizations and they only share collected data through a common server. We will start the analysis of n collaborative UAVs since it is simpler than the noncollaborative cases.

1.3.1 Collaborative UAVs

Each UAV can select a grid line so that no two or more UAVs fly over a same grid line at the same time. This non-overlapping grid lines can be preassigned to them before they depart for wireless data collection. Due to this nonoverlap, the probability of an event being picked up by any one of the UAVs increases,

$$P_{p,n} = \frac{n}{2m}. \tag{1.8}$$

From here, let the subscript n under P represent the number of UAVs used. P_s is independent from the number of UAVs. Therefore,

$$\begin{aligned} P_{d,n} &= P_s P_{p,n} \\ &= \left(1 - \left(1 - \frac{(m-1)d}{L}\right)^2\right) \frac{n}{2m}. \end{aligned} \tag{1.9}$$

Realize that $P_{d,n} = nP_d$, P_d from Eq. (1.4). This means that optimum m^* for multiple collaborative UAVs is the same as Eq. (1.5) and independent from the number of UAVs. As for the maximum detection probability $P_{d,n}^*$, it increases linearly from P_d^* in Eq. (1.6) with increasing number of UAVs. This is a very useful information for the network planner since maximum detection probability reduces as $O(1/k)$ as shown as Fig. 1.2. For example, when the field length L is 100 times larger than the event radius, we can increase P_d to 0.1 by using 10 UAVs instead of one.

1.4 Noncollaborative Independent UAVs

Each UAV is independent and its probability of selecting a grid line is identically distributed for noncollaborative UAVs. Suppose, it is equally likely to select any grid line. Then, an event is picked up if at least one of them selects a correct grid line to detect an event. Therefore,

$$P_{p,n} = 1 - \left(1 - \frac{(m-1)d}{L}\right)^n \tag{1.10}$$

Again, P_s is independent of UAVs so,

$$\begin{aligned} P_{d,n} =& P_s P_{p,n} \\ =& \left(1 - \left(1 - \frac{(m-1)d}{L}\right)^2\right)\left(1 - \left(1 - \frac{1}{2m}\right)^n\right). \end{aligned} \tag{1.11}$$

The optimum $m*$ for noncollaborative UAVs can be obtained by solving the derivative of $P_{d,n}$ in Eq. (1.11). However, there is no simple solution for $m*$ since differentiating Eq. (1.11) leaves us with a function of degree n. So, it is numerically computed as shown as Fig. 1.3. Based on the numerical result it seems that there exists only a single maximum detection probability under our constraint $\frac{L}{2d} + 1 > m \geq 2$ which bounds the size of m from above and below.

Notice from Fig. 1.4 that the optimum size $m*$ grows slowly with respect to the increase of n, the number of independent UAVs. Surprisingly, the result shows that this maximum detection probability with multiple noncollaborative UAVs still increase close to constantly with increase of n. Based on these observations, we are able to obtain a good approximate solution for $m*$ which is also simple enough to reveal the relationship between n and $m*$. Let $\widehat{m*}$ represents the approximated optimum size of network with noncollaborative UAVs.

$$\begin{aligned} f(m) =& P_{d,n} \\ g(m) =& \left(1 - \frac{m-1}{k}\right) \\ h(m) =& \left(1 - (g(m))^2\right) \\ f'(m) =& \frac{-n}{m}\left(\frac{(2m-1)^{n-1}}{(2m)^n}\right)(h(m)) + \frac{(2m)^n - (2m-1)^n}{(2m)^n}\frac{2}{k}(g(m)) \\ =& \frac{-n}{m}\left(\frac{(-1)^{n-1}(1+(-2m))^{n-1}}{(2m)^n}\right)(h(m)) \\ &+ \frac{(2m)^n - \left[(-1)^n(1+(-2m))^n\right]}{(2m)^n}\frac{2}{k}(g(m)) \end{aligned}$$

By representing $f'(m)$ as Eq. (1.12), we can rewrite it as a power series of binomial coefficients. Then, since $(-n)^{n+1} = (-n)^{n-1}c$ for $n \geq 2$, we can further simplify it to the final form as shown as below.

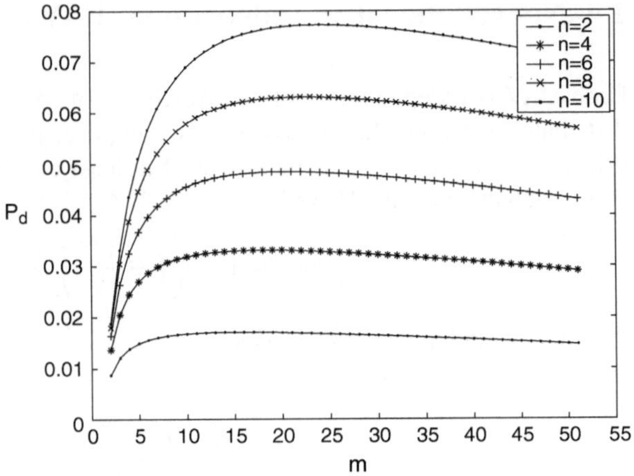

Fig. 1.3 The detection probability of a WSN with multiple noncollaborative UAVs are numerically computed at $k = 100$ for different number of UAVs

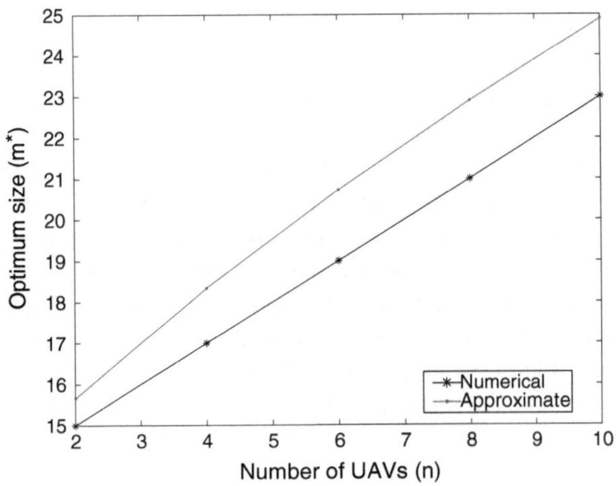

Fig. 1.4 *Linear line* shows the relationship between optimum size ($m^* \times m^*$) of a WSN with multiple noncollaborative UAVs at $k = 100$

$$
\begin{aligned}
&= \frac{-n}{m} \frac{(-1)^{n-1}}{(2m)^n} \left(\sum_{i=0}^{n-1} \binom{n-1}{i} (-2m)^i \right) (h(m)) \\
&\quad + \left\{ (2m)^n - \left[(-1)^n \sum_{i=0}^{n} \binom{n}{i} (-2m)^i \right] \right\} \frac{2(g(m))}{k(2m)^n} \\
&= \frac{-n}{m} \frac{(-1)^{n-1}}{(2m)^n} \left(\sum_{i=0}^{n-1} \binom{n-1}{i} (-2m)^i \right) (h(m)) \\
&\quad + \left[\sum_{i=0}^{n-1} \left(\frac{n}{n-i} \right) \binom{n-1}{i} (-2m)^i \right] \frac{(-1)^{n-1} 2(g(m))}{k(2m)^n}.
\end{aligned} \tag{1.12}
$$

Now by solving this $f'(m) = 0$ we can obtain m^*. However, it is very difficult to solve this directly. So we solve this for the first two terms of the power series with highest degrees of $(n - 1)$ and $(n - 2)$. This will give us the approximate form,

$$-2m^3 + m(3k + nk + n + 1) + (2k - 2nk - n + 1) = 0, \qquad (1.13)$$

which is an cubic function that can be solved using a general formula of roots for a cubic equation or applying Cardano's method we used in Eq. (1.13) which is naturally a discriminant of the cubic equation. Solving it yields to three possible solutions, with only one of them meeting our constraints.

$$\begin{aligned} p &= \frac{3k + nk + n + 1}{-2} \\ q &= \frac{2k - 2nk - n + 1}{-2} \\ u &= \sqrt[3]{-\frac{q}{2} + \sqrt{\frac{q^2}{4} + \frac{p^3}{27}}} \\ \widehat{m^*} &= u - \frac{p}{3u} \end{aligned} \qquad (1.14)$$

This approximated solution can be used as a good rule of thumb for designing an efficient WSN covering widely expanded field with noncollaborative UAVs. This result is plotted in Fig. 1.4 along with optimum size $m*$ obtained numerically. Interestingly, optimum size $m*$ with multiple noncollaborative UAVs increase linearly with an increase in n and approximate solution $\widehat{m^*}$ also closely follows a linear relation. In practice, this result reveals that increasing the number of noncollaborative UAVs will gain constant increase in the maximum detection probability, but also increases the number of sensors to sense the event. But fortunately, the increase of sensors is only linear.

1.5 Conclusion

In this paper, we have investigated the partially controlled deployment of wireless sensors with UAVs as mobile base stations for widely expanded fields.

We have shown that there exists an optimum planning for a given number of UAVs which can maximize the detection probability while maintaining the size of grid network reasonably small. Our analysis shows that the relationship between optimum size of grid and the number of available UAVs for collaborative UAVs is independent of each other. But when the UAVs are noncollaborative, their relationship is almost linear.

Our main contribution is that we have provided a practical, yet an efficient method of covering the vast geographical regions while elongating the lifetime of network. We have also provides a single equation governing the relationship between the optimum size of the grid and the number of available UAVs based on UAVs data collection strategy: collaborative and noncollaborative. These equations can be directly used as a method of designing the optimum planning of UAVs and deployment of wireless sensors networks.

References

1. Cordeiro, C., Agrawal, D.P.: Ad hoc and Sensor Networks. World Scientific Press, Singapore (2006)
2. Zhang, H., Hou, JC.: Is deterministic deployment worse than random deployment for wireless sensor networks? IEEE INFOCOM (2005)
3. Wang, Y., Agrawal, D.P.: Optimizing sensor networks for autonomous unmanned ground vehicles. Optics/Photonics in Security & Defence, pp. 15–18, Cardiff, UK (2008)
4. Clouqueur, T., Phipatanasuphorn, V., Ramanathan, P., Saluja, K.K.: Sensor Deployment Strategy for Target Detection Proceedings. ACM Workshop on Sensor Networks and Applications (WSNA), p. 428 (2002)
5. Tilak, S., Abu-Ghazaleh, N.B., Heinzelman, W.: Infrastructure Tradeoffs for Sensor Networks Proceedings. ACM Workshop on Sensor Networks and Applications (WSNA), p. 498 (2002)
6. Zou, Y., Chakrabarty, K.: Sensor Deployment and Target Localization Based on Virtual Forces Proceedings. IEEE Infocom (2003)
7. Wang, X., Xing, G., Zhang, Y., Lu, C., Pless, R., Gill, C.: Integrated Coverage and Connectivity Configuration in Wireless Sensor Networks, Proceedings. ACM SenSys, p. 289 (2003)
8. Ye, F., Zhong, G., Lu, S., Zhang, L.: PEAS: a Robust Energy Conserving Protocol for Long-Lived Sensor Networks, Proceedings. IEEE ICDCS, p. 287 (2003)
9. Zhang, H., Hou, J.: Maintaining Sensing Coverage and Connectivity in Large Sensor Networks. NSF International Workshop on Theoretical and Algorithmic Aspects of Sensor, Ad Hoc Wireless, and Peer-to-Peer Networks (2004)
10. Wang, W., Srinivasan, V., Chua, K.-C.: Using mobile relays to prolong the lifetime of wireless sensor networks. In MobiCom '05: Proceedings of the 11th annual international conference on Mobile computing and networking. ACM Press, Cologne, Germany (2005)
11. Liu, B., et al.: Mobility improves coverage of sensor networks. In Proceedings of the 6th ACM international Symposium on Mobile Ad Hoc Networking and Computing, MobiHoc '05. Urbana-Champaign, ACM Press, IL, USA (2005)
12. Shah, R.C., et al.: Data MULEs: modeling a three-tier architecture for sparse sensor networks. In: Proceedings of the IEEE Workshop on Sensor Network Protocols and Applications (2003)
13. Bisnik, M., Abouzeid, A.A., Isler, A.A.: Stochastic event capture using mobile sensors subject to a quality metric. IEEE Trans Rob. **23**(4), 676–692 (2007)
14. Leoncini, M., Resta, G., Santi, P.: Analysis of a wireless sensor dropping problem in wide-area environmental monitoring. In: Fourth International Symposium on Information Processing in Sensor Networks. IPSN 2005, pp. 239–245, 15 April 2005
15. Mathai, A.M.: An Introduction to Geometrical Probability: Distributional Aspects with Applications, 1st edn. CRC, Boca Raton (1999)

Chapter 2
A Novel Cross-Layer Architecture for Video Streaming Over MANET

Priya Jumnani and Mukesh Zaveri

Abstract Video streaming is one of the prominent applications of Mobile Ad Hoc Network. Due to high rate requirements, less resource, and severe delay constraints, maintaining real-time media traffic such as audio and video in presence of dynamic network topology is difficult. To cope with this, we propose a cross-layer design (CLD) to optimize the overall performance of video streaming services over MANET. Our CLD jointly controls the video transmission rate, delay constraint, congestion, and resource constraint in order to maximize the received video quality and the performance of the network. When compared with conventional techniques in MANET, our algorithm results in average end-to-end delay, average energy consumption, and the packet loss is considerably reduced with increase in high throughput and good delivery ratio.

Keywords Cross-layer design • MANET • Rate adaptation • Link failure • Congestion control • Energy efficient • MPEG

2.1 Introduction

A Mobile Ad hoc Network is a wireless network consisting of mobile devices that self-configure to form a network without the aid of any established infrastructure. With the rapidly growing popularity of video streaming applications over MANET, the demand for resources efficiency and robustness in the network increases. Due to heterogeneity, dynamism, and unpredictability of

P. Jumnani (✉) · M. Zaveri
Department of Computer Engineering, Sardar Vallabhbhai National Institute of Technology, Surat, India
e-mail: p11co009@coed.svnit.ac.in

M. Zaveri
e-mail: mazaveri@coed.svnit.ac.in

R. Maringanti et al. (eds.), *Proceedings of Ninth International Conference on Wireless Communication and Sensor Networks*, Lecture Notes in Electrical Engineering 299,
DOI: 10.1007/978-81-322-1823-4_2, © Springer India 2014

wireless ad hoc network, QoS demands in terms of delay limitation, bandwidth requirement, and loss tolerance are difficult to guarantee. To cope with this, many cross-layer approaches have been proposed that take into account the specific mobile characteristics.

In cross-layer architecture, layers exchange information and jointly optimize in order to improve the overall performance [1]. Cross-layer design (CLD) exploits layer dependencies and therefore allows us to propagate required parameters throughout the protocol stack [2, 3]. It allows us to make better use of network resources by optimizing across the boundaries of traditional network layers. Hence, it is especially well suited to video streaming application over MANET where the characteristics vary over time. There are many cross-layer designs for different optimization purpose for multimedia communication [4, 5]. Different cross-layer designs focus on different optimization purposes and different QoS metrics, which are delay, priority handling, resource constraint, etc.

While previous work often succeeded in showing the benefits of applying CLD, it often lacked providing solution for handling multiple issues to perform the QoS video transmission over ad hoc network [6, 7]. The cross-layer designs provide individual solution for flow control, admission control, link failure, routing overhead, power conservation, energy minimization, and congestion control. There is no complete and combined solution for the above issues for wireless multimedia applications. In this paper, we propose a cross-layer design framework to provide a combined solution for rate adaptation, link failure management, congestion control, and energy optimization. To achieve high performance under varying conditions, nodes need to adapt their transmission rate dynamically. For addressing the link failure problem, the received signal strength from the physical layer can help to determine the link quality. Links with low signal strength are discarded from the route selection. Routing overhead is improved by minimizing the frequency of recomputed routes by determining whether the packet loss was the result of congestion or node failure causing to compute a new route. For congestion control, the queue length of the nodes can be estimated and notified to the network layer. This estimation of the buffer can be utilized by the network layer and accordingly the optimal path is selected. To achieve energy efficiency, energy metric is used to estimate the remaining energy of the node. This information is then used by network layer to select the energy efficient route.

The paper is organized as follows. Section 2.2 presents the related work done. Section 2.3 presents a detailed description of our proposed architecture. Section 2.4 presents the simulation results and the conclusions are given in Sect. 2.5.

2.2 Related Work

There are numerous research works going on in this area. Rate adaptation strategy falls under two categories, statistics-based and channel quality-based approaches [8]. The [9] approach estimates the channel quality based on successful data transmissions or failures. Using this estimation, it adjusts the data transmission mode in steps. An easy way to obtain the necessary information on wireless channel conditions is to maintain statistics about the transmitted data like the Frame Error Rate/Bit Error Rate (FER/BER), or retry ratio, or the achieved short-term/long-term average throughput. The channel quality-based approach estimates the channel quality based on the measured SNR or SSI instead of the statistics and adjusts the data transmission mode by using the predefined threshold lookup table. Channel-based schemes [10] are further classified into sender based, receiver based, and hybrid adaptation techniques.

Several rate adaptation techniques have been proposed in all cases. Among the existing, some are discussed here. Auto Rate Fallback based on frame loss ratio adjusts the rate based on the number of consecutive successful transmissions. But in ARF [11], frequent collision problems occur. To overcome the collision problem in the ARF scheme, the collision-aware rate adaptation scheme (CARA) is proposed. The key concept of the CARA scheme is that the sender combines adaptively the RTS/CTS exchange with the clear channel assessment (CCA) functionality to differentiate frame collisions from frame transmission failures caused by channel errors. However, a fundamental limit of statistics-based approaches is that they classify channel conditions as either "good or bad." This binary decision provides the direction to adapt the transmission mode, but does not suffice to select the appropriate mode immediately. This leads to a slow step-by-step accommodation to large changes in channel conditions and introduces the risk of oscillation during stable channel conditions. Unlike, the statistics-based approaches, channel quality-based approaches can directly measure the channel conditions and adjust the data transmission mode. An example is the receiver-based auto rate (RBAR) [12] scheme which performs rate adaptation at the receiver instead of the sender.

Link failure management schemes are generally based on the physical layer parameter in order to predict the quality of the link. One of the techniques proposed make use of signal strength as a metric to predict link quality at the network layer. With the incorporation of link layer and transport layer, the proposed Link Adaptive Transport Protocol [13] which provides a systemic way of controlling transport layer offered load for multimedia streaming applications, based on the degree of medium contention information received from the network. In [14] is proposed a cross-layer design that uses a metric known as link residual time (LRT) that is computed based on the received power observed at the physical layer. The value of LRT can be used to estimate the longevity of the link and denotes the remaining time for which the link can be used for packet transmission. LRT values can be used in higher layers to make better decisions for hand-off, scheduling, and routing packets.

The research on congestion for MANET is still ongoing and there is a need for new techniques. Congestion aware routing technique has been proposed by using several metrics such as MAC layer utilization, contention and channel interference, queue length, mobility parameter, etc., to increase throughput of the network. Some of the protocols are discussed here. Congestion Adaptive Routing Protocol (CRP) [15] uses additional paths, called bypass, for bypassing the potential congestion area to the first noncongested node on the primary route. For increasing the performance of network, Congestion aware routing plus Rate Adaptation [16] is proposed which uses two metrics to measure congestion information; average MAC layer utilization and Queue length. To handle delay constraint, Congestion Aware Routing protocol is designed for Mobile ad hoc networks (CARM) [17] based on metrics that incorporate data rates, MAC overhead, and buffer delay to control congestion. For load balancing a new protocol [18], Congestion Aware Scheduling Algorithm for MANET (CARE), is proposed, which decreases the arrival rate at the congested node and balances the load among network.

Energy is the important issue to be addressed because the Ad hoc network has limited resource. Several researches have proposed many algorithms [19] for handling the limited resource in an efficient manner. Energy Efficient MANET Routing Protocols are mainly divided into two categories based on how they minimize the active communication energy, i.e., transmission control approach [20] and load distribution approach [21]. For minimizing energy during inactivity sleep/power-down mode approach has been proposed [22]. Energy efficient routing protocols based on transmission power control find the best route that minimizes the total transmission power between a source–destination pair. The specific goal of the load distribution approach is to balance the energy usage of all mobile nodes by selecting a route with underutilized nodes rather than the shortest route. The sleep/power-down mode approach focuses on inactive time of communication.

2.3 Proposed Cross-Layer Design

2.3.1 Overview

In this paper, we propose a novel cross-layer architecture named DSR-CE, whose aim is to provide a combined solution for rate adaptation, link failure management and reduced routing overhead, congestion control and energy efficient for dealing with issues like packet loss, delay, power constraint, and QoS. In DSR-CE, Physical, MAC, and Network layers cooperate with each other to fulfill the task of QoS provision to video streaming application. Our main contribution is to provide the solution for the following critical issues for video streaming over mobile ad hoc network.

We first integrate rate adaptation module with link failure module named DSR-RL, which makes use of received signal strength parameter for cross-layer optimization. In rate adaptation module, transmission rate selection selects data rate in the MAC layer based on the channel estimation information from physical layer. To achieve high performance under varying conditions, nodes need to adapt their transmission rate dynamically. To address the link failure problem, the received signal strength from the physical layer can help to determine the link quality. The

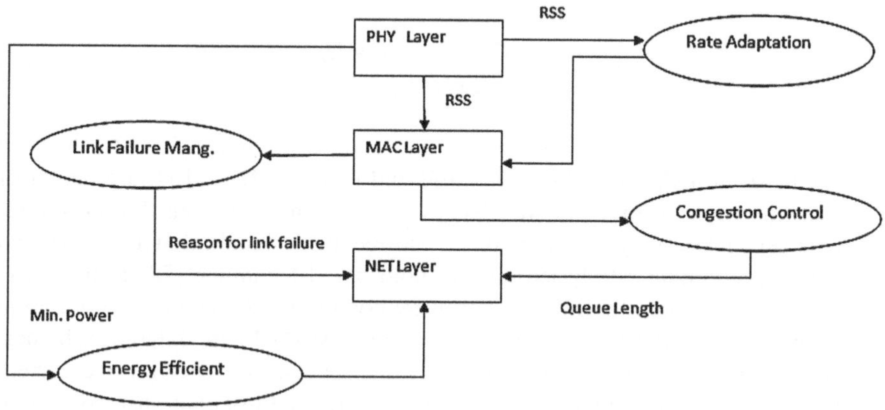

Fig. 2.1 Cross-layer architecture

links with low signal strength are discarded from the route selection. Routing overhead is improved by minimizing the frequency of recomputed routes by determining whether the packet loss was the result of congestion or node failure that caused to compute a new route.

Then, further we modified the design and proposed a new architecture, i.e., DSR-CE which contains solution for congestion and resource constraint also in MANET. For congestion control, the queue length of the nodes can be estimated and notified to the network layer. This estimation of the buffer can be utilized by the network layer and accordingly the optimal path is selected. In energy efficiency module, energy metric is used to estimate the remaining energy of the node from Physical layer. This information is then used by network layer to select energy efficient route. More detailed description of the module is given below. The architecture is shown in Fig. 2.1.

2.3.2 Rate Adaptation

Video applications usually require low error rates. Since video over wireless ad hoc network is already compressed at the available bit rate, errors can ruin the video quality at the receiver. To maintain a constant low error rate, the bit rate of the video bit stream must be adapted to the available bit rate. This would require that physical layer channel information be made available at the MAC layer. Our approach to solve the problem of fluctuating available bit rate makes use of channel state information at the sender to determine at which rate the bit stream is to transfer. The receiver estimates the signal strength of transmission channel using channel model simulation at the physical layer. The parameter used for channel estimation in our scheme is received as power indication P_r, which is calculated as

$$P_r = \frac{P_t * G_t * G_r * H_t * H_r * \lambda^2}{(4 * \pi * d)^2 * L} \tag{2.1}$$

where, P_t and P_r are signal power at receiver and transmitter, G_t and G_r are gain for a signal to a node from the transmitter and receiver, H_t and H_r are height of the transmitter and receiver, is wavelength, d is distance between the transmitter and receiver and L is system loss. Then the received signal strength is mapped to a transmission data rate based on threshold-based technique [23] at the MAC layer. In this approach, the rate is chosen by comparing the channel information based on series of thresholds related to available M-QAM modulation schemes. The receiver sends the determine bit rate to the sender. On receiving the current rate from the receiver, the physical layer at the sender adjusts its transmission rate accordingly. Other neighbor nodes that hear the packet will update the information in their network allocation vector (NAV) and hold their transmission until current transmission gets completed.

2.3.3 Link Failure Management

In ad hoc wireless networks, we deal with nodes that have different power capabilities; hence, there is a considerable likelihood to transmit with different power levels. The link asymmetry arises when a node with high power transmits to a lower power node, and the high power-node cannot sense an ongoing transmission triggered by the low power node. Thereby, the hidden terminal problem is exacerbated, which provokes more false link failures, increasing the time that route discovery process is launched. Normal DSR interprets a link failure (in MAC layer) as a broken link, even when caused by congestion at receiver. The received signal strength of neighboring nodes can be used to detect the reason for lost packets, distinguishing between congestion and broken links due to mobility, because in broken links due to mobility, the receiver is not reachable. The reason for unsuccessful communication is sent to the routing layer. The routing layer should interpret that communication to destination was not possible, not because of a broken link but rather congestion; therefore, route maintenance is not needed. If that is not the reason delivered to the routing layer, a route maintenance process is required. We calculate and use the average signals of nodes to stop retransmitting packets when destination is not reachable because it moved away. The proposed scheme will make DSR distinguish between both situations, avoiding the route error process when the link error at MAC layer is due to congestion and not due to mobility of nodes causing broken links. When MAC layer is not able to communicate to a neighboring node, MAC layer informs to the routing layer not only that there was a problem, but also includes if the neighboring node is still reachable.

2.3.4 Congestion Aware and Energy Efficient Routing

The energy consumption can be reduced by designing routing protocols of these select routes with less energy consumption for end-to-end packet transmission. Congestion aware routing protocol is designed such that it reduces collision and selects the route that avoids congested nodes. We design energy efficient routing and congestion control protocol of MANETs named as DSR-CE. It discovers the route based on energy aware metrics and congestion aware metrics. Here, we use two metrics: queue length for congestion and remaining energy of node. The congestion was assumed to be in existence when queue length was near capacity or when battery level fell below a predefined threshold. The proposed protocol works similar to normal DSR protocol, if the current queue length is less than maximum queue length. When a data packet needs to wait in queue for a longer time, there is a possibility for unexpected delay in transmission or dropping of packets. The number of packets in the queue is a metric reflecting the traffic load of the mobile node. A mobile node with more traffic passing through it usually has more packets in its interface queue. Average queue size indicates the node traffic load in the long term [24]. The calculation of the average queue size is updated every T seconds based on the following equation:

$$\text{len} = \beta * \text{qlen} + (1 - \beta) * \text{qlen}_{\text{sample}} \tag{2.2}$$

where qlen denotes the average queue length and $\text{qlen}_{\text{sample}}$ the current queue length which is constant and set to 0.3 in our simulations. Communication is the main source of energy consumption for a mobile node. Nodes consume energy while transmitting data to a desired destination when forwarding data while acting as intermediate nodes between source and destination nodes, or receiving data destined to them. Hence, we calculate energy metric as power consumed with the following equation:

$$E_n = P_{\text{tx}} - P_r + P_{\text{th}} + P_m + P_{\text{over}} \tag{2.3}$$

where E_n denotes the total energy consumed by node, P_{tx} is power consumed in transmission mode, P_r is power consumed in reception mode, P_{th} is receiving power threshold, P_m is power consumed in idle mode, and P_{over} is power consumed in overhearing mode. When an intermediate node receives a route discovery (RREQ) packet, the following steps are undertaken by the algorithm: Node checks the source node's energy level E_s, source node ID, which was advertised in the network, and the destination node ID. It saves all attributes in its route cache. It calculates whether its queue length cost from equation exceeds the threshold value, i.e., the maximum queue length. If the queue length cost is more than the threshold, it will not forward the packet to any other intermediate node and will not participate in the routing process. The busy node simply drops the RREQ packet, hence the discovered route simply omits the congestive nodes. If the queue length cost is less than the threshold, it appends its node ID, energy level and forwards the packet to its neighboring nodes. When destination node receives the RREQ message it will calculate the mean value of all the values of E_n of all the nodes and send an RREP message to the node whose E_n value is nearest to the mean value. If the route discovery broadcast is not the first one (i.e., if it is another communication period, *T*), the node will check its route cache to see if there exist routes to the destination. In case there is one, it chooses the one with shortest hop and with minimum total energy to the destination with congested free nodes.

2.4 Simulation Results

In this section, we illustrate the feasibility and effectiveness of the proposed CLD for video streaming, i.e., DSR-CE (DSR with rate adaptation, link failure management, congestion control, and energy efficient) in the NS-2 simulator and compare it with DSR-RL (DSR with rate adaptation and link failure management) and DSR under varying background traffic load, different frame rates [25], different video sequences, and different mobility scenarios based on various performance metrics,viz., network and QoS metrics. We test the accuracy of design for different network varying environment. For analyzing the dynamic behavior of ad hoc network, static and mobile scenario with different mobility of 1–10 m/s are taken for experiment. We use IEEE 802.11 as MAC layer protocol. It has the functionality to notify the network layer about link breakage. In our simulation, 50, 75, 100 mobile nodes move in a 500 m * 500 m rectangular region for 200 s simulation time. All nodes have the same transmission range of 250 m. The background traffic is simulated using CBR packets with fixed size of 1,024 bytes.

For video traffic, the encoded packet traces of different video sequence of both QCIF format (176 * 144) and CIF (352 * 288) such as Akiyo, Foreman, Mother Daughter, News, Hall, High-way, Bridge, and Football are used [26]. To test the effect of rate adaptation, we vary offered load by changing the CBR packet rate from 10 packets/second to 60 packets/second and video sequence at a frame rate of 10, 20, and 30 fps. An analysis of the performance based on mobility is performed by varying the pause time of the node from 20 s (high mobility) to 100 s (low mobility). The number of nodes is taken as 50 (small area network), 100 (large area network) and the maximum number of connections as 20, 40 with different video sequences (low and high resolution).

2.4.1 Network Level Evaluation

It indicates the efficiency of the tested protocols for indicating perceived performance of network. We analyzed how well our design performs in different networking environments using the following metrics.

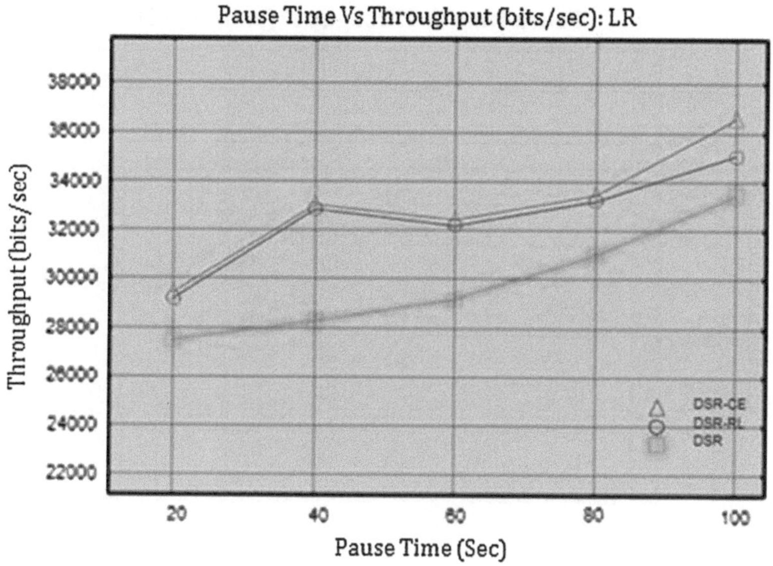

Fig. 2.2 Throughput for network of 50 nodes for low resolution video

1. *Throughput*: In case of high mobility, DSR-CE, DSR-RL increases the overall throughput of the network. The number of link errors in DSR-CE, DSR-RL is decreased compared to normal DSR, because of the high reduction of routing packets, achieving higher overall throughput. In DSR, the length of the route fluctuates between high and optimal values, due to a large amount of route errors and the consequent new routes found. By reducing route errors and congested free route in DSR-CE, the length of the routes becomes optimal and constant (Figs. 2.2, 2.3).
2. *Delivery Ratio*: Packet delivery ratio of DSR is less compared to DSR-CE, DSR-RL because the number of packets dropped is more in DSR. Due to link failure management, less number of packets can be dropped which results in good packet delivery ratio (Figs. 2.4, 2.5).
3. *End-to-End Delay*: The delay of DSR-RL, DSR is almost similar because in case of congestion both schemes wait for a certain period of time before they send the packets that introduce more congestion hence it introduces delay; while in case of DSR-CE, it finds out the congested free path, hence the probability of congestion is reduced (Figs. 2.6, 2.7).
4. *Energy Consumption*: DSR-CE outperforms DSR under different traffic loads, which is mainly due to the benefit of power control in the Network layer. The excess packets inevitably introduce more collisions to the network, wasting more energy. DSR-CE chooses alternative routes, avoiding the heavily burdened nodes and thus alleviating the explosion in average energy consumption (Fig. 2.8).

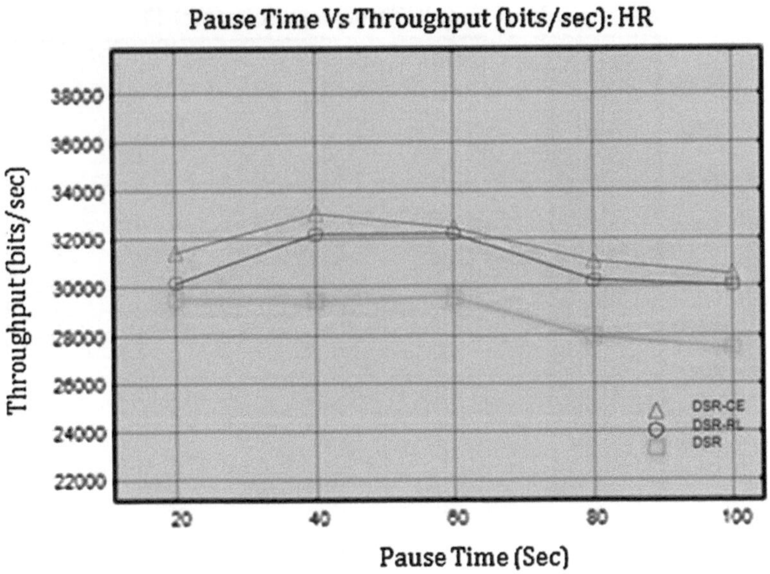

Fig. 2.3 Throughput for network of 50 nodes for high resolution video

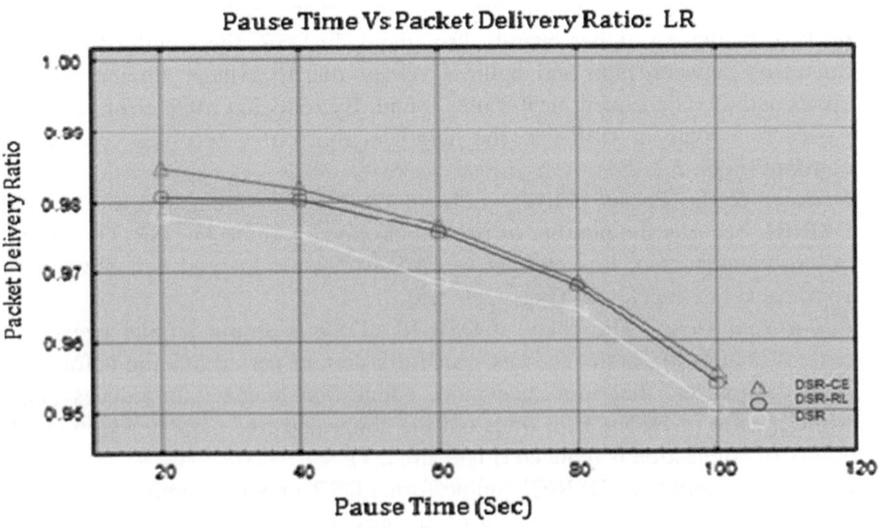

Fig. 2.4 Packet delivery ratio for network of 50 nodes for low resolution

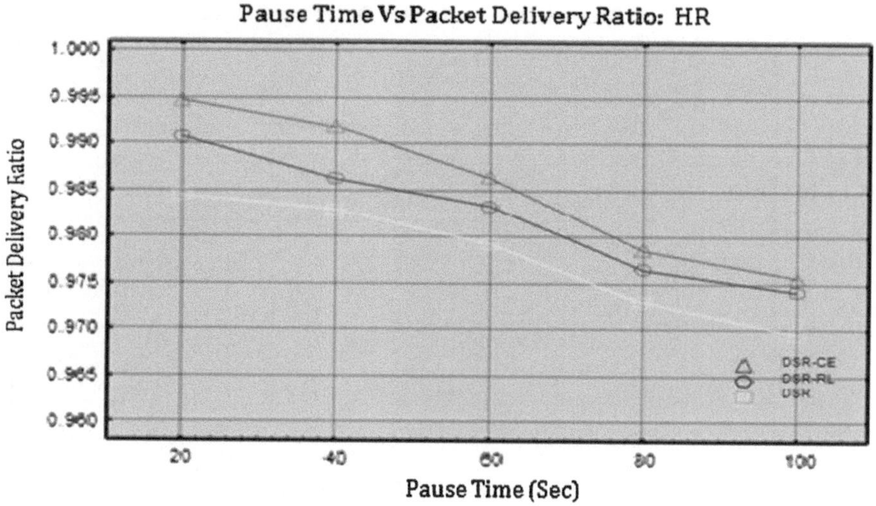

Fig. 2.5 Packet delivery ratio for network of 50 nodes for high resolution

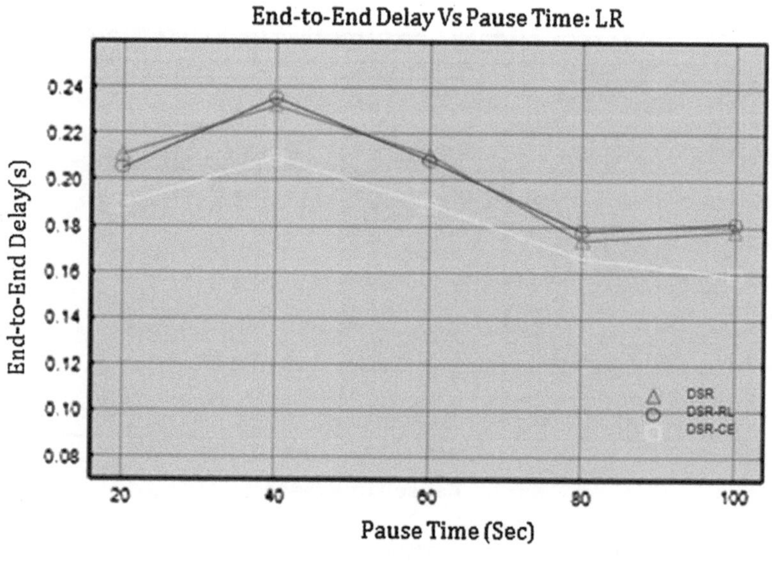

Fig. 2.6 End-to-end delay for network of 50 nodes for low resolution video

2.4.2 Quality of Experience Measurement

Measure of network parameter results in control of resource but for maintaining user satisfaction we need QoS parameter. Quality of experience is the overall acceptability of an application or service, as perceived subjectively by end-users. QoE assessment divided into two different ways: subjective and objective.

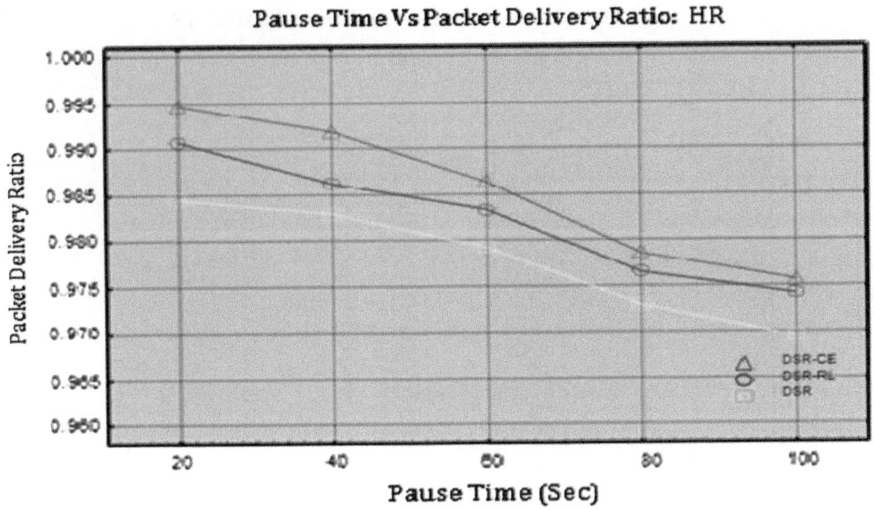

Fig. 2.7 End-to-end delay for network of 50 nodes for high resolution video

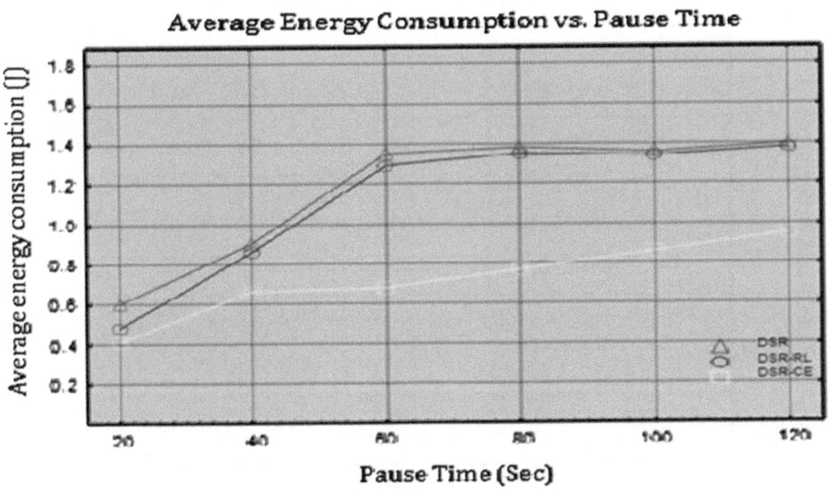

Fig. 2.8 Energy consumption versus pause time

1. *PSNR*: From the result as shown in table, we observe the average PSNR of DSR-CE outperforms DSR-RL by 0.9 dB and DSR by 1.6 dB. We see that, DSR-CE and DSR-RL have lower end-to-end delay compared to DSR, while DSR-CE has the lowest packet loss rate 0.0028. DSR-CE can balance the traffic over multiple paths to avoid the congestion occurs which lead to lowest packet loss probability and improve the robustness of wireless transmission. To quantify the improvement that our algorithm produces on

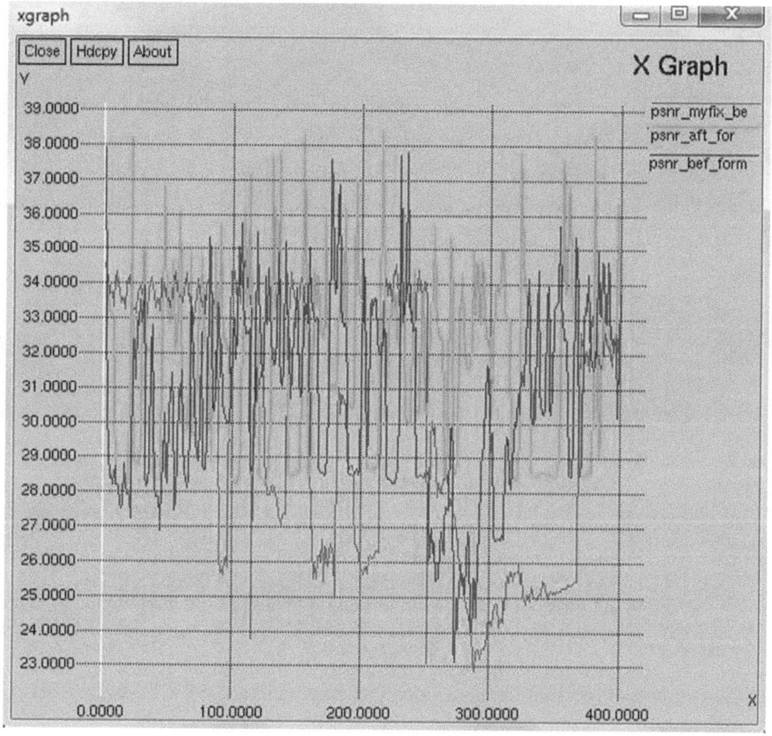

Fig. 2.9 PSNR of video transmission after simulating in DSR-CE, DSR-RL, and DSR

Table 2.1 PSNR, delay and loss rate

	Packet loss rate	Delay (s)	Average PSNR(dB)
DSR-CE	0.0028	0.0548	33.458
DSR-RL	0.0045	0.0879	32.516
DSR	0.076	0.1457	31.795

steaming video, we calculate the PSNR of each frame once before our algorithm has been implemented and once after our algorithm has been implemented. Figure 2.9 shows the PSNR fluctuation received after transmission of video through DSR-CE, DSR-RL, and DSR (Table 2.1).

2. *Subjective Assessment*: To illustrate the perceptual video quality delivered by different approaches, we observe the quality of a video in YUV Viewer which gives results of the distortion and delay occurred during transmission. From Fig. 2.10 it can be seen that quality of video improves when transmitting through DSR-CE, DSR-RL as compared to DSR. DSR introduces distortion and delay in receiving frame which degrades the quality of video. DSR-CE has better performance because it can reduce delay and distortion by avoiding or relieving the congestion.

Fig. 2.10 Screenshots of received foreman video sequence with (**a**) DSR-CE, (**b**) DSR-RL, (**c**) DSR

2.5 Conclusion

Based on the analysis of ongoing efforts, cross-layer design appears to be a suitable approach for future contributions to the framework of wireless network that address emerging issues related to ever-higher performance, energy consumption, and mobility. We have designed a cross-layer design-based architecture to provide a combined solution for link failure management, rate adaptation, congestion control, and energy efficiency. Performance of DSR-CE is compared with DSR-RL and DSR. It increases the QoS of video in terms of PSNR as well as subjective quality. By simulation results, we have shown that the DSR-CE reduce average end-to-end delay, average energy consumption, and packet loss with increase in high throughput and good delivery ratio.

References

1. Srivastana, V., Motani, M.: Cross-layer design: a survey and the road ahead. IEEE Commun. Mag. **43**(12), 112–119 (2005)
2. Shakkottai, S., Rappaport, T., Karlsson, P.: Cross-design for wireless networks. IEEE Commun. Mag. **3**(1), 74–80 (2003)
3. Gavrilovska, L.: Cross-layering approaches in wireless ad hoc networks. Wireless Pers. Commun. **37**(9), 271–290 (2006). (Springer)

4. Kawadia, V., Kumar, P.: A cautionary perspective on cross layer design. IEEE Wireless Commun. **12**(3), 234–235 (2005)
5. Rao, Santhosha, Shama, Kumara: Cross layer protocols for multimedia transmission in wireless networks. Int. J. Comput. Sci. Eng. Surv. (IJCSES) **3**(3), 113–119 (2012)
6. Bouras, Hristos, Gkamas, Apostolos, Kioumourtzis, Georgios: Challenges in cross layer design for multimedia transmission over wireless networks. IEEE Sig. Process. **11**(3), 301–317 (2010)
7. Lindeberg, M., Kristiansen, S., Plagemann, T., Goebe, V.: Challenges and techniques for video streaming over mobile ad hoc networks. Multimedia Syst. **17**(6), 51–72 (2011). (Springer)
8. Ramachandran, B., Shanmugavel, S.: Received signal strength-based cross-layer designs for mobile ad hoc networks. IETE Tech. Rev. **25**(4), 192–200 (2009)
9. Delgado, G.D., Fria, V.C., Igartua, M.A.: ViStA-XL: a cross-layer design for video-streaming over ad hoc networks. IEEE **6**(8), 15–19 (2008)
10. Biaz, S., Wu, S.: Rate adaptation algorithms for IEEE 802.11 networks: a survey and comparison. IEEE **1**(6), 124–187 (2008)
11. Kim, J., Kim, S., Choi, S., Qiao, D.: CARA: collision-aware rate adaptation for IEEE 802.11 WLANs. IEEE INFOCOM **41**(4), 22–27 (2006)
12. Wong, S., Yang, H., Luand, S., Bharghavan, V.: Robust rate adaptation for 802.11 wireless networks. Mobile Commun. **27**(3), 146–157 (2006)
13. Pavon, J., Choi, S.: Link adaptation strategy for IEEE 802.11 WLAN via received signal strength measurement. ICC **31**(5), 1108–1123 (2003)
14. Chen, X., Zhai, H., Wang, J., Fang, Y.: TCP performance over mobile ad hoc networks. Electr. Comput. Eng. Can. J. **29**(1), 129–134 (2004)
15. Y.-C. Hu, D.B. Johnson (2004) Exploiting congestion information in network and higher layer protocols in multihop wireless ad hoc networks. In: Proceedings of the the 24th International Conference on Distributed Computing Systems, IEEE, vol. 31, no. 8, pp. 301–310. (2004)
16. Chen, X., Jones, H.M., Jayalath, A.D.S.: Congestion-aware routing protocol for mobile ad hoc networks. Veh. Technol. Conf. **31**(6), 21–25 (2007)
17. Wu, W., Zhang, Z., Sha, X., Qin, D.: A study of congestion-aware routing protocols for wireless ad-hoc network. IEEE **8**(5), 39–47 (2009)
18. Chen, Z., Ge, Z., Zhao, M.: Congestion aware scheduling algorithm for MANET. Nat. Nat. Sci. Found. CHINA **16**(41), 41–47 (2005)
19. Goldsmith, A.J., Wicker, S.B.: Design challenges for energy-constrained ad hoc wireless networks. IEEE Wireless Commun. **9**(4), 8–27 (2002)
20. Ray, N.K., Turuk, A.K., Energy efficient technique for wireless ad hoc network. In: Proceedings of International Joint Conference on Information and Communication Technology, vol. 23, no. 4, pp. 105–111. (2010)
21. Tarique, M., Tepe, K., Naserian, M.: Energy saving dynamic source routing for adhoc wireless networks. In: Proceedings of 10th International Conference on Modeling and Optimization in Mobile Ad Hoc and Wireless Networks, vol. 13, no. 3, pp. 21–31. (2005)
22. Sivasankar, P., Chellappan, C., Balaji, S.: Performance of energy efficient routing protocol for MANET. Int. J. Comput. Appl. **28**(8), 1–6 (2011)
23. Chauhan, R.K., Chopra, Ashish: Power optimization in mobile ad hoc network. Global J. Comput. Sci. Technol. **10**(4), 92–96 (2010)
24. Goldsmith, J., Chua, S.G.: Variable-rate variable power M-QAM for fading channels. IEEE Trans. Commun. **45**(2), 1218–1230 (1997)
25. Lin, C.-H., C-H, Ke, Shieh, C.-K., Chilamkurti, N.K., Zeadally, S.: A novel cross-layer architecture for MPEG-4 video stream over IEEE 802.11e wireless network. Spec. Issue Int. J. Telecommun. Syst. **23**(4), 211–221 (2008)
26. MPEG-4 and H.263 video traces for network performance evaluation. http://www.tkn.tu-berlin.de/research/trace/trace.html

Chapter 3
RPL-SCSP: A Network-MAC Cross-Layer Design for Wireless Sensor Networks

Raja Ben Abdessalem and Nabil Tabbane

Abstract In this paper, we propose a network-MAC cross-layer design protocol called Routing Protocol for Low-power and Lossy Networks-Sleep Collect and Send Protocol (RPL-SCSP) for wireless sensor networks (WSNs) that optimizes both the end-to-end delay and the energy consumption. RPL-SCSP improves the RPL that supports real-time quality of service (QoS), to ensure a reduced end-to-end delay for urgent data by selecting the parent taking into account the number of packets present in the queue of each potential parent node. Our proposed cross-layer protocol has proven its efficiency through simulation results obtained from COOJA simulator under Contiki operating system, in comparison with the basic media access control (MAC) and network protocols.

Keywords Wireless sensor network • QoS • Energy consumption • End-to-end delay • Optimization • MAC protocols • SCSP • RPL • Contiki • COOJA

3.1 Introduction

A wireless sensor networks (WSN) is deployed over a geographical area to monitor physical phenomena. It has variant applications such as medical monitoring, homeland security, and military applications. WSNs typically use limited life

R. Ben Abdessalem (✉) · N. Tabbane
Mediatron Laboratory, Higher School of Communication of Tunis, University of Carthage, Tunis, Tunisia
e-mail: benabdesslem.raja@supcom.rnu.tn

N. Tabbane
e-mail: nabil.tabbane@supcom.rnu.tn

R. Ben Abdessalem
ESPRIT, Institute of Engineering and Technology, Tunis, Tunisia

R. Maringanti et al. (eds.), *Proceedings of Ninth International Conference on Wireless Communication and Sensor Networks*, Lecture Notes in Electrical Engineering 299,
DOI: 10.1007/978-81-322-1823-4_3, © Springer India 2014

battery that cannot be easily replaced or recharged because sensors are deployed in hostile environments. To have a long lifetime, the individual nodes should reduce the power consumption. Thus, several power-saving mechanisms and media access control (MAC) protocols [1] have been designed and studied to reduce the power consumption by switching the radio off as often as possible such as SCSP [2]. SCSP works according to the paradigm "sleep, collect, and send". It does not require synchronization between routers. In fact, to make the receiver aware of the transmission, the first message to transmit during the send period is preceded by a preamble. In a communication stack, the MAC and routing protocols cooperate together in order to ensure the communication between neighbors. The RPL is emerging as a Proposed Standard "Request For Comment" RFC6550 in the Internet Engineering Task Force (IETF) [3, 4].

In this work, we propose RPL-SCSP protocol, a network-MAC cross-layer design that guarantees a rapid transmission for urgent data while minimizing the energy consumption. It presents, in the network layer, a modified version of the RPL protocol. In the MAC layer, we use a modified version of the SCSP protocol that is tolerant to reduce the energy consumption by switching its node between sleep and active modes. Furthermore, SCSP dynamically adapts the active and sleep periods (SPs) according to the received traffic in order to achieve high throughput rates when an urgent event occurs. Moreover, urgent events and critical information require a rapid transmission time (end-to-end delay). Therefore, a node should choose a less loaded parent node. Accordingly, we define a threshold to the received packets number per each queue. MAC layer provides the list of neighbor routers to the network layer in order to offer to the node multiple forwarding choices. Hence, MAC protocol has the possibility to change the next hop router if the number of packets in its parent's queue reaches the threshold. In addition to the optimization of the end-to-end delay, RPL-SCSP optimizes the energy consumption by adapting a new mechanism that reduces the active state periods. Indeed, RPL-SCSP turns the radio off, if no packet is queued during the collect period of the SCSP MAC protocol. We implement our cross-layer in the Contiki Operating System [5, 6] and we compare it with the conventional protocols (SCSP and RPL) by simulating different networks through the COOJA simulator.

The remainder of the paper is organized as follows. Section 3.2 presents the related work. Section 3.3 describes the cross-layer design. In Sect. 3.4, we present the simulation results of our proposed cross-layer. Finally, Sect. 3.5 concludes the paper.

3.2 Related Work

Much research has been addressed to minimize the energy consumption such as MAC protocol designs, which use duty cycling mechanisms, in which nodes periodically switch between sleep and active modes. S-MAC [7], B-MAC [8], and IEEE 802.15.4 [9] are examples of energy efficient MAC protocols for WSNs. S-MAC has a synchronization overhead that maintains a common sleep schedule between

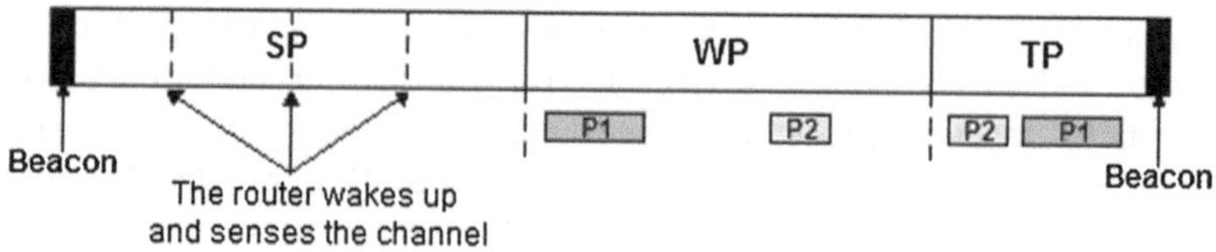

Fig. 3.1 Superframe structure

nodes, whereas B-MAC is an asynchronous protocol that uses an adaptive preamble scheme to achieve low-power communication. In fact, the router periodically wakes up and checks the channel activity. When a preamble is received, the router stays awake to receive the incoming packet. IEEE 802.15.4 standard provides a physical and a MAC layer for low-cost and low-rate wireless networks. This MAC protocol can operate on both beacon enabled and non-beacon modes. In its beacon mode, the IEEE 802.15.4 uses a superframe structure which is defined as the time between two successive beacon transmissions and composed of an active and a SP.

Nefzi et al. [2] proposed a MAC protocol called SCSP that benefits from the advantages of S-MAC and B-MAC to provide further enhancements. SCSP uses SP and wake up period such as in S-MAC protocol and it operates like B-MAC during the SPs. The idea of SCSP is that a router sleeps for a period of time, wakes up, and collects data from other routers and then sends them into a burst during a period of time that we call a transmission period (TP). After joining the network, a router starts sending beacon frames to provide useful information to its neighbors. The time between two successive beacon frames is called superframe. The superframe is divided into three periods; the SP during it, the router enters into sleep mode, the waiting period (WP) during it the router becomes active, it collects data from its neighbors, and the TP during it the router starts transmitting all packets queued in its buffer in a single burst. The superframe structure is illustrated in Fig. 3.1.

When the WP finishes, the router transmits a preamble, then it starts transmitting all packets queued in its buffer during TP. At the end of the TP, the router sends the next beacon and starts another cycle.

Many routing protocols are suitable for duty cycle MAC protocols like Zigbee [10] tree routing (ZTR) and RPL [3, 4]. Zigbee is a wireless standard based on IEEE 802.15.4 standard for Personal Area Networks. It uses a low data transfer rate to monitor application. Nefzi et al. [2] used a modified version of the ZTR (m-ZTR) with SCSP which is more tolerant to node failure. M-ZTR does not require route discovery. In fact, end devices in ZTR always send their data to their father, whereas, in m-ZTR, due to the beacon frame send by the SCSP protocol at the beginning of the superframe, end devices send their data to the first received beacon sender. Accordingly, routing decision in m-ZTR uses neighborhood table. Hence, the flexibility of the selection of next hop router maintains the network connectivity before the received traffic falls, which make m-ZTR more tolerant to node failures than ZTR. Thus, in comparison with IEEE 802.15.4 MAC, the cross-layer in [2] extends the network lifetime.

The routing protocol RPL is designed by the working group Routing Over Low-power and Lossy networks (ROLL) for Low-power and Lossy Networks (LLNs) where constrained devices were interconnected by wireless and wired lossy links. RPL is a distance vector protocol that builds a Directed Acyclic Graph (DAG) where paths are constructed from each node in the network to the DAG root (Sink). DAGs offer redundant paths which are required for LLNs. To build the DAG, RPL uses a discovery mechanism based on ICMPv6 messages. When forming the topology, the sink and the nodes send a DAG Information Object (DIO) message to advertise information about the DAG such as the DAGID and the object function (OF). To let arbitrary nodes communicate with each other, RPL defines a destination advertisement object (DAO) message to advertise the prefix reachability toward the children and to propagate the information along the DAG in the up direction to populate the routing tables of the ancestor nodes. Much research has been addressed to enhance RPL such as in [11] which investigated the use of a MAC protocol called receiver-based MAC (RB-MAC). Such approach enhances the end-to-end delay and increases the energy efficiency relying on the current channel condition. In fact, RB-MAC protocol reduces the number of retransmission by electing the neighbor which has a strong link to the sending node.

3.3 QoS Protocol Based on Cross-Layer Design

The QoS is a critical component of the overall architecture. Some data such as the critical alarms have a real-time requirement and the networking infrastructure should guarantee an optimized end-to-end delays. So, for delay sensitive applications, it is important to limit this parameter. RPL-SCSP is a cross-layer protocol that uses in its MAC layer a modified version of the SCSP and in its network layer, a modified version of the RPL, holds the promise to achieve higher performances in terms of energy consumption and end-to-end delay.

3.3.1 Routing Protocol

RPL finds the best path according to OFs. It finds the path that minimizes the path expected transmission count (ETX), where the path ETX is defined as the sum of the ETX for all the traversed links [4]. The ETX metric characterizes the average number of the required transmitted packets to ensure a successful transmission. The ETX is consequently tightly coupled with the throughput along a path. Therefore, the OF chooses the parent with the minimum rank (ETX), as illustrated in Algorithm 1.

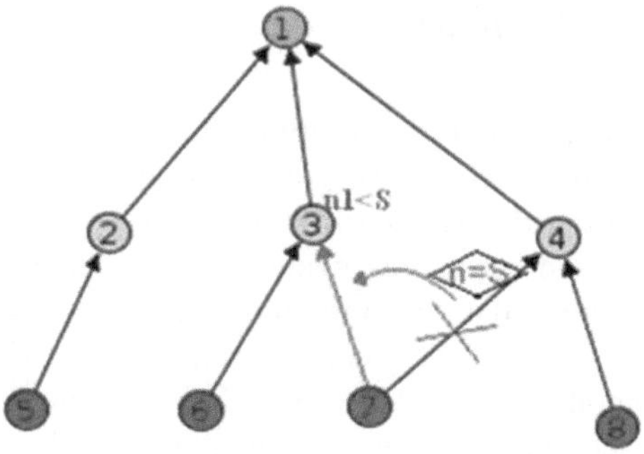

Fig. 3.2 Improvement of end-to-end delay

Algorithm 1:

```
if (p1_rank < p2_rank)
   { return p1 ; }
else
   { return p2 ; }
```

Moreover, the end-to-end delay depends on the number of packets in the queue during WP "nqpacket." The more the number of packets in the queue increases, the more the end-to-end delay will be longer. Therefore, we define a threshold S that represents the maximum number of packets in the queue during the WP (maximum of "nqpacket"). Thus, other than ETX, the OF depends on "nqpacket." The new mechanism works as follows. During the WP, each router node sends the number of packets in its queue to its neighbors. In this way, other than ETX, the parent should be selected according to the parameter "nqpacket." To minimize the end-to-end delay, this number should be lower than the threshold S (see Algorithm 2). In addition, when "nqpacket" reaches the threshold S, the node should abandon its parent. The mechanism is illustrated in Fig. 3.2.

Algorithm 2:

```
if (p1_rank < p2_rank)
   { if ((p1→nqpacket < S) || (p2→nqpacket >= S))
     { return p1 ; }
   else if (p2→nqpacket < S)
     { return p2 ; }
   }
```

However, if the number of the received packets in the queue of all the prospective parents exceeds S, the node selects the parent with a minimal rank (ETX).

3.3.2 Channel Access Protocol

To provide a long lifetime, the power consumption should be low for the system. Thus, to reduce the power consumption, SCSP switch the radio off (sleep mode) during the SP. During this period, the router goes to sleep; it periodically wakes up and senses the channel. If the channel is busy, it remains active as far as it receives data from the medium. Otherwise, it goes back again to sleep. When the SP expires, the router becomes active. However, it does not transmit any packet upon their arrival in the WP. When the WP expires, the router stays in the active state, and transmits all packets queued in its buffer. However, the radio is always active during WP even if the queue is empty during this period. To improve the energy consumption, SCSP should switch the radio off during WP if the queue is empty and wakes up periodically to sense the channel. Hence, it is important to provide an empty queue during the WP. To provide this, the node should choose a parent that has already packets in its queue.

3.3.3 QoS Support

In order to ensure the required QoS, our cross-layer provides a compromise between the energy efficiency and the end-to-end delay. Thus, the proposed mechanism of the OF is improved (see Algorithm 3). This new mechanism aims to enhance the energy consumption by choosing the parent that has already packets in its queue ("nqpacket" $= 0$) and to enhance the end-delay by choosing a parent with "nqpacket" less than S. Accordingly, the node maintains its parent only if "nqpacket" is between 1 and S.

Otherwise ("nqpacket" $= 0$ or "nqpacket" $>$ S), the node selects another parent such that the number of received packets in its queue ("nqpacket") is equal to a value between 1 and S$-$1. In the case where queues of all prospective parents are empty, the node selects the best parents with a minimal rank (ETX).

Algorithm 3:

```
if (p1_rank < p2_rank)
  { if(((p1 → nqpacket < S)&&(p1 → nqpacket > 0))||(p2 → nqpacket >= S))
   { return p1; }
   else if ((p2→nqpacket < S) && (p2→nqpacket > 0))
   { return p2; }
   else { return p1; }
  }
```

3.4 Simulation Results

In this section, we present a simulation study of our cross-layer (RPL-SCSP) which we implemented in the Contiki operating system [5, 12] and we simulated in Contiki COOJA network simulator. We consider the network illustrated

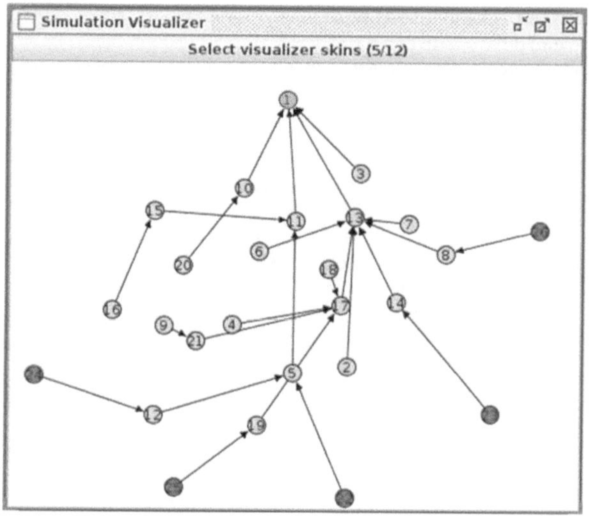

Fig. 3.3 Simulation scenario

Table 3.1 Simulation parameters

Parameters	Value
Simulation time (min)	30
Maximum transmission radius (m)	50
Clear channel assessment (ms)	8
RDC channel check rate (Hz)	8

in Fig. 3.3, composed of 20 routers (yellow circles in the figure), 5 nodes (pink circles), and a sink node (green circle). The sink node (the root) is always active.

We analyze the energy consumption and end-to-end delay with different traffic loads and we compare RPL-SCSP with the basic protocols RPL and SCSP. Table 3.1 summarizes the general simulations parameters.

3.4.1 End-to-End Delay

As first simulation, we analyze the end-to-end delay as a function of the traffic load. Furthermore, the end-to-end delay depends on the chosen threshold S. The maximum number of packets in the queue is fixed to 4. By this way, S could be chosen between 1 and 3. We varied the traffic and then measured the end-to-end delay with different values of the threshold S. We get the best end-to-end delay with a threshold S equal to 3. This first simulation, illustrated in Fig. 3.4, shows that RPL-SCSP provides better results than the basic protocols. In fact, due to the new algorithm, the packet does not expect much time to be transmitted for a heavier traffic load.

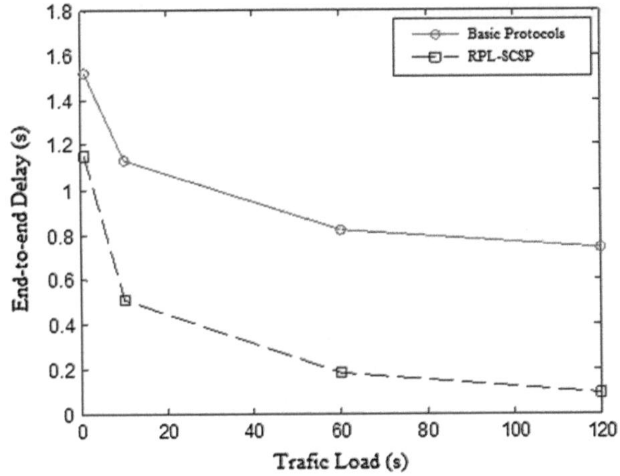

Fig. 3.4 Comparison of end to end delays between basic protocols and RPL-SCSP (S = 3)

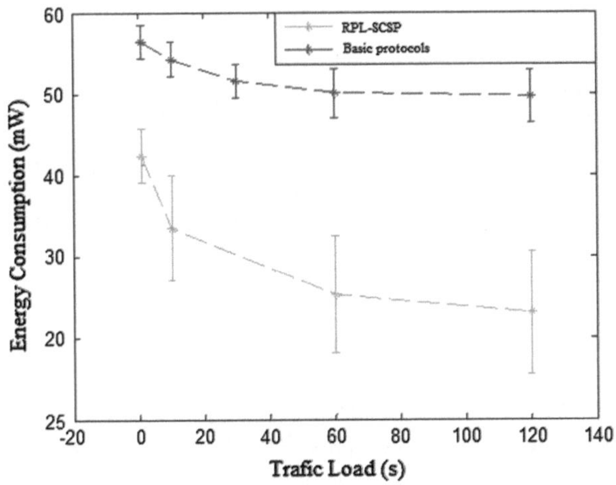

Fig. 3.5 Comparison of energy consumption with different traffic load between RPL-SCSP and basic protocols

3.4.2 Energy Consumption

In this section, we analyze the energy consumption in the network, illustrated in Fig. 3.3, by varying the traffic load during the simulated time. Figure 3.5 shows that RPL-SCSP outperforms the basic protocols SCSP and RPL. Indeed, our cross-layer achieves better results in terms of energy consumption. Moreover, when using the basic protocols, the radio is always on during the WP even if there is no packet in the queue during this period. When the traffic load decreases, the energy consumption also decreases. This is explained by the fact that the node rarely switch its radio on during the WP and the TP when the traffic load is low.

3.5 Conclusion

In this paper, we proposed a new approach to improve the QoS in wireless sensor networks using a network-MAC cross-layer design RPL-SCSP. RPL-SCSP provides a reduced end-to-end delay in order to guarantee a rapid transmission of urgent data, by proposing a new algorithm to select the best parent according to the queue load. RPL-SCSP aims also to optimize the energy consumption in order to extend the network lifetime, by switching the node into inactive state each time the queue is empty. Thus, this proposed cross-layer supports QoS by offering a good compromise between the energy consumption and the end-to-end delay. The simulation results have shown that, in comparison with the basic protocols, the RPL-SCSP cross-layer design reduce the end-to-end delay and extend the network lifetime by optimizing the energy consumption.

References

1. Huang, P., Xiao, L., Soltani, S., Mutka, M., Xi, N.: The evolution of MAC protocols in wireless sensor networks: a survey. Commun. Surv. Tutorials **99**, 1–20 (2012). IEEE
2. Nefzi, B., Cruz-Sanchez, H., Song, Y.-Q.: Scsp: an energy efficient network-MAC cross-layer design for wireless sensor networks. In: Proceeding of IEEE 34th Conference on Local Computer Networks 2009. LCN 2009, October 2009, pp. 1061–1068
3. Winter, T., Thubert, P., Brandt, A., Hui, J., Kelsey, R., Levis, P., Pister, K., Struik, R., Vasseur, J., Alexander, R.: RPL: IPv6 Routing Protocol for Low-Power and Lossy Networks. RFC 6550 (Proposed Standard) (March 2012)
4. Vasseur, J., Dunkels, A.: Interconnecting Smart Objects with IP: The Next Internet. Morgan Kaufmann, Burlington (2010)
5. Dunkels, A., Gronvall, B., Voigt, T.: Contiki-a lightweight and flexible operating system for tiny networked sensors. In: Proceeding of 29th Annual IEEE International Conference on Local Computer Networks, November 2004, pp. 455–462
6. Tsiftes, N., Eriksson, J., Dunkels, A.: Low-power wireless ipv6 routing with contikirpl. In: Proceedings of the 9th ACM/IEEE International Conference on Information Processing in Sensor Networks 2010, pp. 406 –407
7. Ye, W., Heidemann, J., Estrin, D.: Medium access control with coordinated adaptive sleeping for wireless sensor networks. IEEE/ACM Trans. Netw. **12**(3), 493–506 (2004)
8. Polastre, J., Hill, J., Culler, D.: Versatile low power media access for wireless sensor networks. In: Proceedings of 2nd International Conference on Embedded Networked Sensor Systems (Sensys'04), pp. 95–107. Baltimore, MD, USA (2004)
9. IEEE-TG15.4.: Part 15.4: Wireless medium access control (MAC) and Physical Layer (PHY) specifications for low-rate Wireless Personal Area Networks (LR-WPANs). IEEE standard for information technology (2003)
10. 473-489 ZigBee Specification Document 053474r17. http://www.zigbee.org
11. Akhavan, M., Watteyne, T., Aghvami, A.: Enhancing the performance of rpl using a receiver-based MAC protocol in lossy wsns In: Proceeding of 18th International Conference on Telecommunications (ICT) 2011, May 2011, pp. 191–194
12. Heddeghem, W.V.: Cross-layer link estimation for Contiki-based wireless sensor networks. Master's thesis, Vrije Universiteit Brussel, August 2009

Chapter 4
Congestion Avoidance and Lifetime Maximization in Wireless Sensor Networks Using a Mobile Sink

Sagar Motdhare and C. G. Dethe

Abstract Congestion severely affects the performance of a wireless sensor network in two aspects: increased data loss and reduced lifetime. This paper addresses these problems by introducing a mobile sink for congestion avoidance and lifetime maximization in wireless sensor networks (WSN). Also, in the proposed scheme data only has to travel a limited number of hops to reach the mobile sink, which helps to improve the energy consumption of the sensor nodes. We have considered various parameters like packet delay, packet loss, and throughput for evaluation. Through simulation, we show effectiveness of the proposed scheme in terms of congestion avoidance and increased lifetime of the wireless sensor network.

Keywords Congestion avoidance • Lifetime maximization • Mobile sink • Wireless sensor network

4.1 Introduction

The phenomenon of congestion can be observed in different types of wired and wireless networks even in the presence of strong routing algorithms. Congestion in wireless sensor networks (WSN) mainly occurs because of two reasons—when multiple nodes want to transmit data through the same channel at a time or when the routing node fails to forward the received data to the next routing nodes because of the *out-of-sight problem*.

S. Motdhare (✉) · C. G. Dethe
PIET, Nagpur, India
e-mail: sagar_motdhare@yahoo.com

C. G. Dethe
e-mail: cgdethe@gmail.com

R. Maringanti et al. (eds.), *Proceedings of Ninth International Conference on Wireless Communication and Sensor Networks*, Lecture Notes in Electrical Engineering 299,
DOI: 10.1007/978-81-322-1823-4_4, © Springer India 2014

Applications of WSNs in the areas of environment and habitat monitoring require the sensor nodes to periodically collect and route data toward a sink. Also, it is known that each sensor node can only be equipped with a limited amount of storage, so if at any given routing node the data collection rate dominates the data forwarding rate congestion starts to build up at this node. Such type of congestion and data loss normally occurs at the nodes located in the vicinity of a static sink. Data loss at these nodes occurs due to the fact that at any given point of time a sink can only communicate with one or a limited number of sensor nodes [1, 2].

To mitigate this *static sink problem*, new strategies have been developed by exploiting the mobility of a sink to better balance the problem of congestion and the energy consumption among the sensors. That is, the mobile sink traverses the monitoring region and rest at some locations to collect sensed data. It has been demonstrated that sink mobility is a blessing rather than a curse to network performance including the network lifetime, packet delay, packet loss, and throughput.

The lifetime of the sensor network is another important aspect in environmental and habitat monitoring-based applications of the WSN. Lifetime of a WSN can be defined as the time interval between the deployment of the sensor field and the time when the first sensor node fails due to complete energy dissipation [1, 3, 4, 2].

The main contribution of this paper is that we have done the analysis of the effect of mobile sink in reducing congestion and increasing lifetime of the sensor network by considering various parameters. The same parameters are considered for the sensor network using static sink. It has been shown that, sensor network with a mobile sink performs better to reduce congestion as compared to sensor network with a static sink.

The rest of the paper is organized as follows: Sect. 4.2 summarizes related work, Sect. 4.3 presents the sink mobility model, simulation setup and analysis of results are given in Sects. 4.4 and 4.5, respectively and Sect. 4.6 concludes the paper.

4.2 Related Works

In this section, a summary of currently available congestion avoidance techniques and their removal is discussed. It also elaborates the recent work done on investigating the use of mobile sink in WSNs.

4.2.1 Congestion Avoidance and Control Techniques

In WSN, there are mainly two reasons that result in congestion—(1) the packet arrival rate exceeding the packet service rate. This occurs at sensor nodes, which are in the vicinity of the sink as they usually carry more upstream traffic. (2) Contention, interference, and bit error rate on a link also results in congestion [4, 2]. Congestion has a direct impact on energy efficiency and application QoS. It

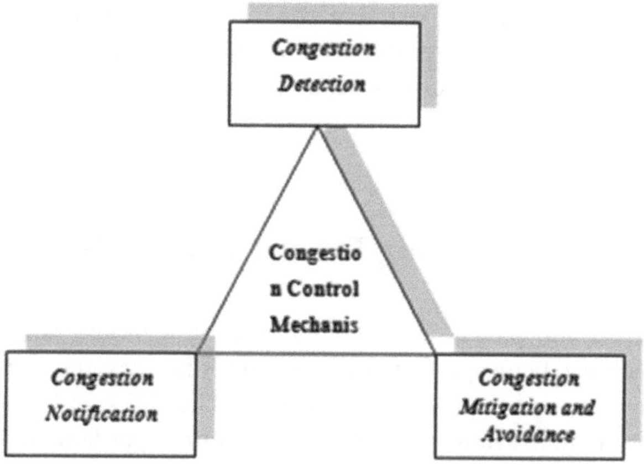

Fig. 4.1 Congestion avoidance and control techniques

can cause buffer overflow, packet loss, and can also degrade link utilization. Thus, congestion in WSN must be efficiently controlled (Fig. 4.1).

Chen [5] classified the congestion control techniques in WSN into two groups: congestion avoidance and congestion control. The previous part focuses on strategies to avoid congestion from taking place and the latter works on removing congestion when it has occurred. From an implementation perspective these techniques can be categorized into three groups: data aggregation techniques, multi hop/path routing techniques, and flow control techniques.

Data aggregation techniques focus on utilizing spatial or temporal correlation between sensed data to reduce its quantity and hence prevent congestion.

Multi hop/path routing techniques utilize the dense deployment of the sensor nodes to remove congestion from WSNs.

Flow control techniques try to control the amount of data that is flowing on the routing path to avoid congestion using various strategies.

In addition to above-mentioned schemes Wang [6] proposed a node priority-based congestion control scheme for WSN. Their scheme is based on the supposition that the nodes located in a WSN have diverse bandwidth and wireless media control requirement for data transmission. Therefore, a node priority index can be generated on the basis of packet inter arrival time and service time at each node. With the help of this index, sensor nodes having heavy data traffic can be assigned more access to the transmission media than the nodes with less traffic. However, extra overhead is involved in this scheme for maintaining the priority index of the sensor nodes.

In [7], the author has proposed a metrics called depth of congestion (DC) to detect congestion. When DC exceeds the predefined threshold value it intimates, that, congestion has occurred. In this issue, it necessitates to develop an effective and efficient congestion control protocol to avoid the network from entering the congestion collapse state. This is done by implementing Hop by Hop congestion control protocol.

Investigation of the given schemes has shown that although they avoid or eliminate congestion, but they fail to avoid data and energy loss during this process [1, 4, 8].

To summarize, existing congestion control schemes lead to data loss. In order to overcome this problem, a WSN with a mobile sink is presented in this paper. Now will we discuss various types of sink mobility models for data routing to a mobile sink.

4.3 Data Routing Toward a Mobile Sink

Over the past few years, the use of a mobile sink has increased in WSNs to achieve better performance, in particular for balanced utilization of the sensor field energy and to prolong the lifetime. The sink can follow three basic types of mobility patterns in a WSN: random mobility, predictable/fixed path mobility, and controlled mobility.

4.3.1 Random Mobility

Chatzigiannakis [9] used random sink mobility in their schemes. In case of random mobility, the sink follows a random path in the sensor field and implements a pull strategy for data collection from the sensor nodes. On the other hand, with random sink mobility it is not possible to guarantee data collection from all the sensor nodes positioned in a WSN.

4.3.2 Predictable/Fixed Path Mobility

Luo [10] have tried to find a mobility strategy for the sink that can lead to the most energy efficient utilization of the sensor field. The longest lifetime for the sensor network can only be achieved if the mobility route of the sink is along the *periphery* of the sensor field. Increased data latency and packet loss are the major problems that happen due to the sink mobility in WSN. One potential drawback of their scheme is that whenever the sink moves, routing paths need to be updated.

4.3.3 Controlled Mobility

Use of controlled sink mobility is also analyzed in WSNs for increasing the lifetime. Jayaraman [11]. Controlled sink mobility-based schemes are a good option if compact data latency is required, but they are less cost-effective than random/ fixed path mobility. If data latency is permissible, then the best routing strategy that incurs minimum data loss due to sink mobility and also provides maximum

Table 4.1 Simulation parameters

Parameter	Value
Channel type	Wireless channel
MAC protocol	IEEE 802.11
Frequency/bandwidth	2.4 GHz/250 kbps
Interface queue type	Queue/Droptail
Traffic type	CBR
Antenna model	Omni directional antenna
Sink speed	2 m/s
No. of nodes	22
Maximum packets in interface queue	50
Packet size	512 bytes
Routing protocol	AODV
Simulation time	80 s
Simulation area	1,500 × 900

lifetime of the sensor network with minimum cost is obtained if the sink follows a discrete mobility pattern along the boundary of the sensor field.

To summarize, if data latency is permissible, then the best routing strategy that incurs minimum data loss due to sink mobility and also provides maximum lifetime of sensor network with minimum cost is obtained if the sink follows a discrete mobility pattern along the boundary of the sensor field [1, 12].

4.4 Simulation Environment

In the presented model, we consider 22 nodes out of which 21 nodes are WSN nodes and 1 node is a sink node. The wireless sensor nodes are uniformly but randomly deployed. Nodes are responsible for sensing and reporting their readings at constant time to the sink. Sensor nodes are grouped into clusters and each cluster has a head node that is responsible for forwarding the received data from neighboring client nodes toward sink.

We have considered the following parameters for wireless sensor network with a static sink as well as WSN with a mobile sink. The following metrics have been selected for evaluating the effect of mobile sink in order to reduce congestion and increase the lifetime of the sensor network. The comparison is shown in the 6 section (Table 4.1).

4.4.1 Packet Loss

It is the ratio of number of packets lost in the network to number of packets generated by the sensing nodes.

$$\text{Packet Loss} = \frac{\text{No. of Packets Lost in the Network}}{\text{NO. of Packets Generated by the sensing Nodes}}$$

Packet loss in the network should be as low as possible to increase the reliability of the network [13].

4.4.2 Packet Delay

Packet delay is the total latency experienced by a packet to traverse the network from the source to the destination. At the network layer, the end-to-end packet latency is the sum of processing delay, packet, transmission delay, queuing delay, and propagation delay. The end-to-end delay of a path is the sum of the node delay at each node plus the link delay at each link on the path [13, 14].

4.4.3 Throughput

It is the average rate of successful packet delivery over a Sensor Network. The throughput is usually measured in data packets per second.

$$\text{Throughput} = \frac{\sum \text{(Packet Size)}}{\text{(Packet Arrival Time)} - \text{(Packet Start Time)}}.$$

A high network throughput indicates a small error rate for packet transmission and the low level for contention in the network [14, 13].

4.4.4 Lifetime

Lifetime is the important aspect in WSN. It is defined as the interval between the deployment of the sensor field and the time when the first sensor node fails due to complete energy dissipation [1].

Figure 4.2 shows the simulation Environment which typically consists of 22 nodes out of which 21 nodes are wireless sensor nodes and one node is a sink node. The same environment is considered for both static as well as mobile sink scenarios and the above parameters are considered for the comparison. Later we present the graphical analysis of both the scenarios.

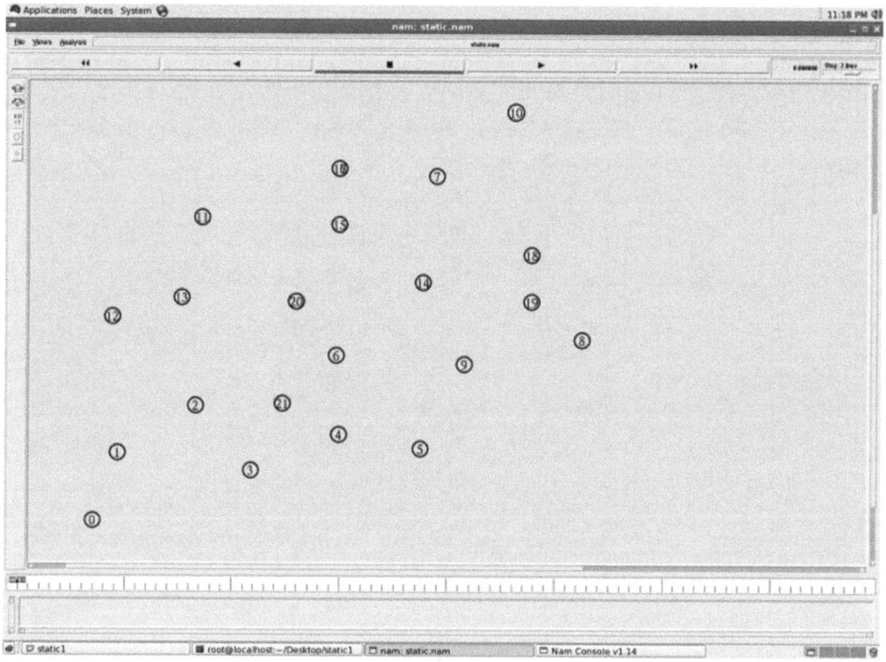

Fig. 4.2 Simulation environment

4.5 Result Analysis

Besides running independently, both the simulations are capable of obtaining the graphs which shows the packet lost in the network, packet delay, and throughput. It can be seen from the simulation results that in a WSN with a static sink, the nodes which are located in the vicinity of a static sink, get congested as it is heavily burdened due to the fact that it is responsible for carrying the traffic of the neighboring nodes. Once congestion starts to build up in the network, the battery of the congested node drains out quickly, which ultimately results in node failure.

From Figs. 4.3 and 4.4, which depicts a static sink WSN, it is very much clear that the node located near to the sink gets congested and die out quickly due to complete loss of energy. It may be noted that congestion also takes place due to many to one transmission, which also results in heavy packet loss and decreased throughput.

If the same scenario is considered, but in place of static sink, a mobile sink is introduced, then the data has to travel limited number of hops to reach the sink which helps to improve the energy consumption of the sensor nodes. Also, there will be no funneling effect taking place as there is no extra burden on nodes as shown in Figs. 4.5 and 4.6.

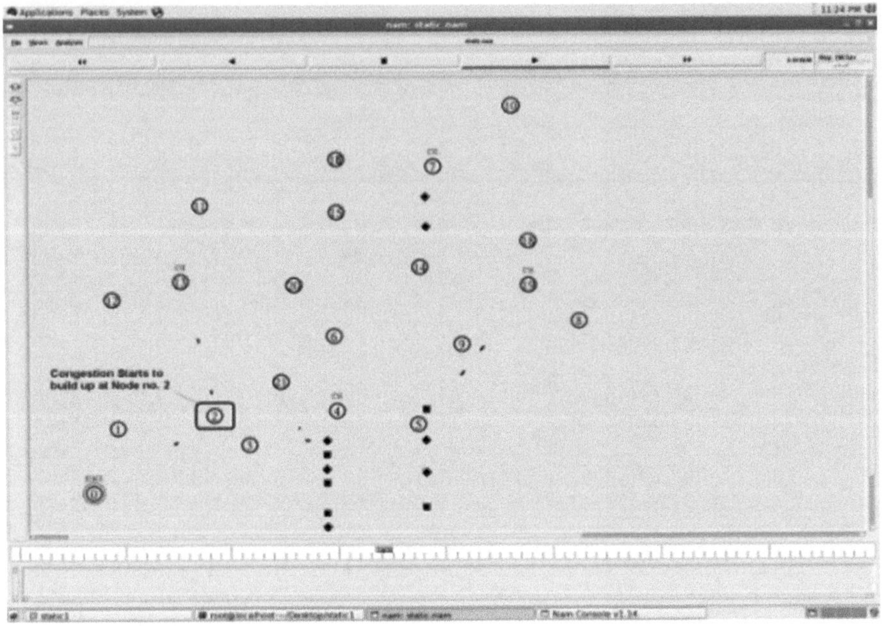

Fig. 4.3 Congestion starts to build up at node 2 which is located in the vicinity of a static sink and it fails due to complete loss of energy

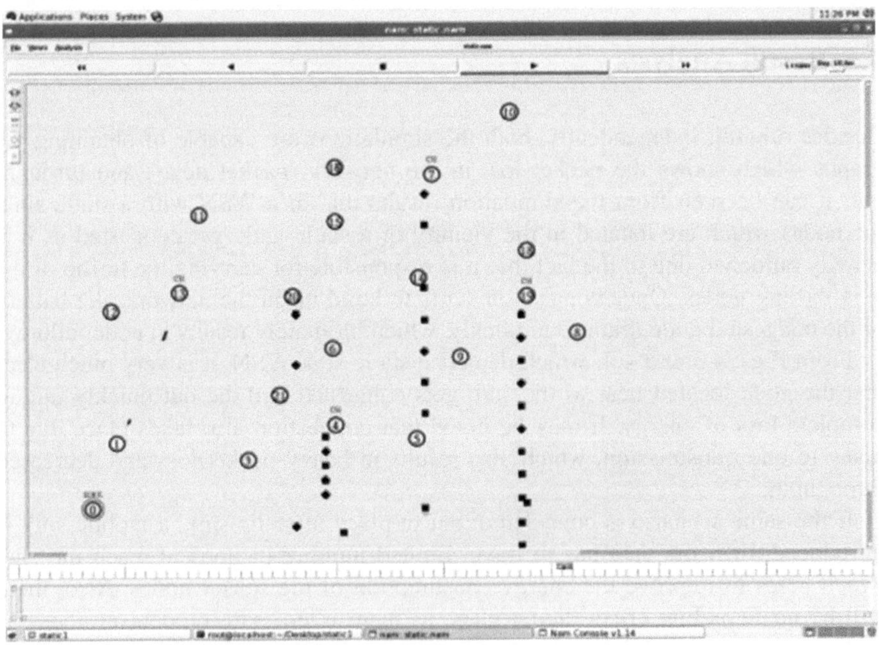

Fig. 4.4 Congestion starts to build up at node 2 which is located in the vicinity of a static sink and it fails due to complete loss of energy

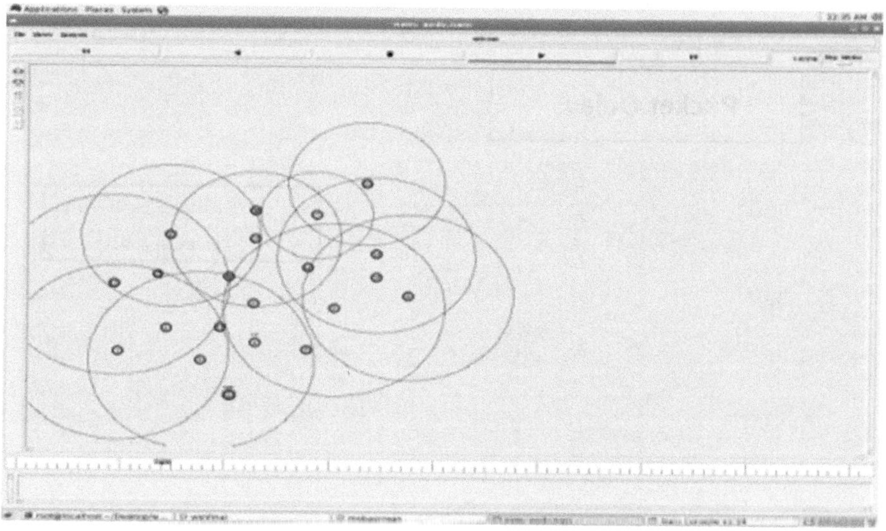

Fig. 4.5 Simulation environment with mobile sink and cluster head no. 4 and 13 transmitting data toward the mobile sink

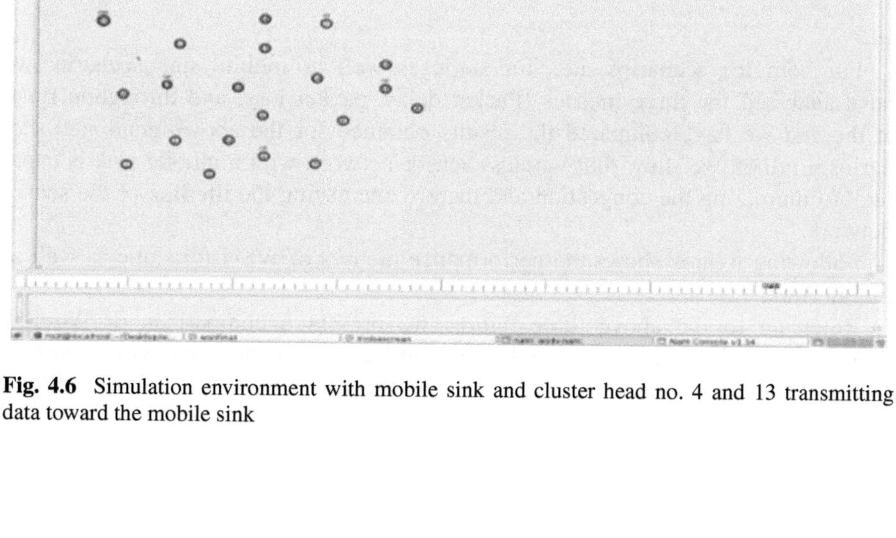

Fig. 4.6 Simulation environment with mobile sink and cluster head no. 4 and 13 transmitting data toward the mobile sink

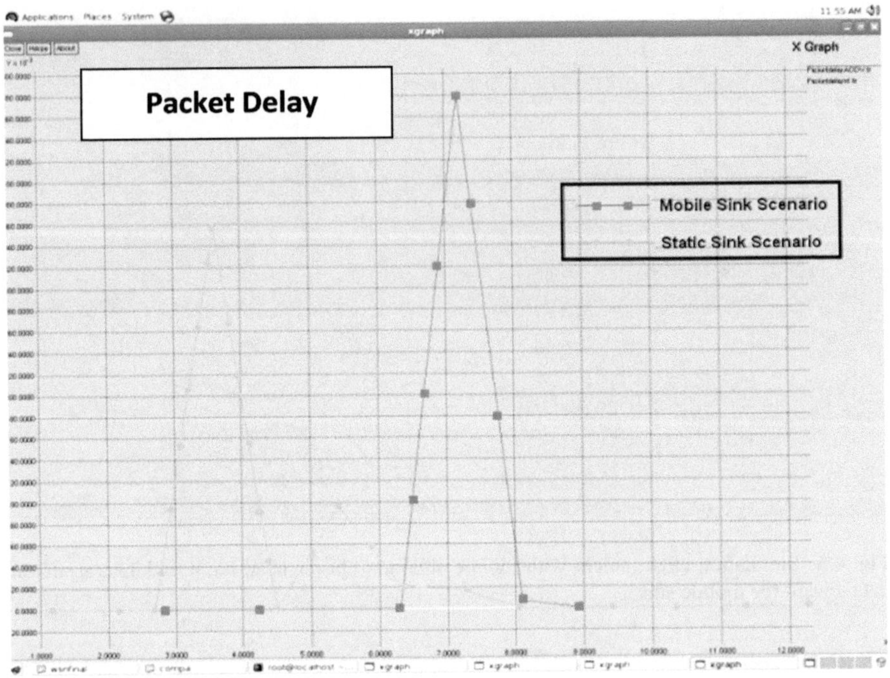

Fig. 4.7 Graph of time versus packets transmitted

Table 4.2 Comparison of parameters

Parameter	Static scenario	Mobile scenario
Packet delay	More	Less
Packet loss	Heavy	Reduced
Throughput	Less	Increased
Lifetime	Reduced	Increased

For both the scenarios, i.e., for static as well as mobile sink scenario, we have analyzed the three metrics (Packet delay, packet loss, and throughput) and at the end we have compared the results obtained for the above-mentioned scenarios. Further, we show that wireless sensor network with a mobile sink is capable of minimizing the congestion and thereby increasing the lifetime of the sensor network.

Following figures shows the performance metrics of WSN for static as well as mobile sink.

Referring to the above three figures we present a comparison of Wireless Sensor Network with a static sink and wireless sensor network with a mobile sink by considering the three parameters as shown in the Figs. 4.7, 4.8, and 4.9 (Table 4.2).

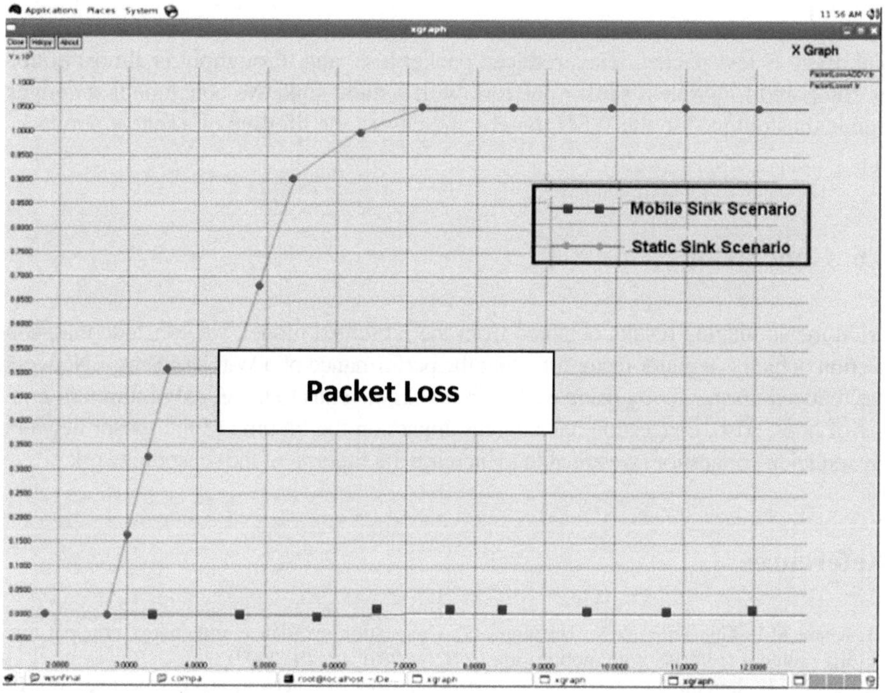

Fig. 4.8 Graph of time versus packet loss

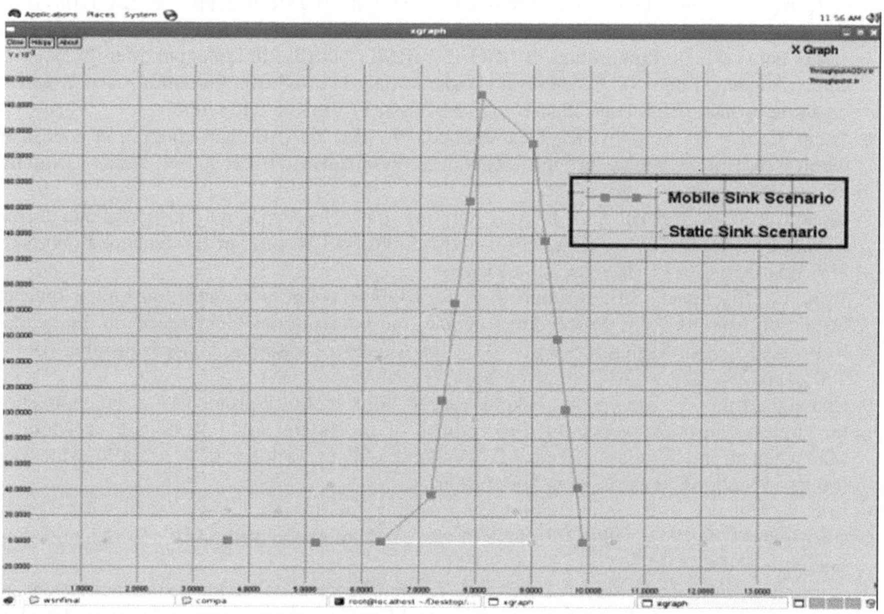

Fig. 4.9 Graph of time versus packets delivered successfully

From the table it is clear that in case of wireless sensor network with a mobile sink there is less packet delay, reduced packet loss, and throughput is almost double as compared to wireless sensor network with a static sink. We conclude that congestion is thus reduced in the WSN thereby increasing the lifetime of a sensor network.

4.6 Conclusions

From the simulation results obtained from the presented model, we conclude that, congestion, which is a major factor affecting the performance of a Wireless Sensor Network, has reduced drastically by using a *Mobile sink* as data has to travel only minimum number of hops. Also, congestion has a direct impact on the lifetime of the sensor network. By reducing congestion, we are able to increase the lifetime of the sensor network.

References

1. Khan, M.I., Gansterer, W.N., Haringm, G.: Congestion avoidance and energy efficient routing protocol for WSN with mobile sink. J. Netw. **2**(6), 42–49 (2007)
2. Zheng, J., Jamalipour, A.: Wireless sensor networks: a networking perspective, WILEY Publications, New york. ISBN: 978-0-470-16763-2
3. Khan, M.I., Gansterer, W.N., Haring, G.: In-network storage model for data persistence under congestion in wireless sensor network. In: Proceedings of the First International Conference on Complex, Intelligent and Software Intensive Systems (CISIS'07), pp 221–228. (2007)
4. Sagar, S., Motdhare, C., Dethe, G.: Study of various congestion contol protocols in wireless sensor networks. In: Proceedings of NCOAT-NIRMITI-2013, PIET, Nagpur, Mar-2013
5. Chen, Shigang, Yang, Na: Congestion avoidance based on lightweight buffer management in sensor networks. IEEE Trans. Parallel Distrib. Syst. **17**(9), 934–946 (2006)
6. Wang, C., Li, B., Sohraby, K., Daneshmand, M., Hu, Y.: Upstream congestion control in wireless sensor networks through cross-layer optimization. IEEE J. Sel. Areas Commun. **25**(4) (2007)
7. Chakravarthi, R., Gomathy, C.: Hop-by-hop rate control technique for congestion due to concurrent transmission in wireless sensor network. World Comput Sci Inf Technol J (WCSIT) **1**(8), 351–356 (2011). (ISSN: 2221-0741)
8. Wan, C., Eisenman, S.B., Campbell, A.T.: CODA: congestion detection and avoidance in sensor networks. In: Proceedings of the 1st International Conference on Embedded Networked Sensor Systems, SenSys '03, (Los Angeles, California, USA, November 05–07, 2003), pp. 266–279. ACM Press, New York, NY (2003)
9. Chatzigiannakis, I., Kinalis, A., Nikoletseas, S.: Sink mobility protocols for data collection in wireless sensor networks. In: Proceedings of the International Workshop on Mobility Management and Wireless Access (Terromolinos, Spain, October 02–02, 2006), MobiWac '06, pp. 52–59. ACM Press, New York (2006)
10. Luo, J., Hubaux, J.-P.: Joint mobility and routing for lifetime elongation in wireless sensor networks. In: Proceedings of the 24th Annual Conference of the IEEE Communications Societies (INFOCOM'05), FL, USA (2005)
11. Jayaraman, P.P., Zaslavsky, A., Delsing, J.: Sensor data collection using heterogeneous mobile devices. In: ICPS'07: IEEE International Conference on Pervasive Services, Istanbul, Turkey, 15–20 July 2007

12. Guo, J.: Sink mobility schemes in wireless sensor networks for network lifetime extension University of Windsor, Windsor, guo1d@uwindsor.ca
13. Birla, J., Basant, S.: Performance Metrics in Ad-hoc Network. BRCM, Bahal
14. Gowrishankar, S., Basavaraju, T.G., Sarkar, S.K.: Simulation based analysis of mobile sink speed in WSN. In: Proceedings of the World Congress on Engineering and Computer Science, vol. I (2010)

Chapter 5
Providing Stable Routes in Mobile Ad Hoc Networks

Arka Prokash Mazumdar, Adhar Surange and Ashok Singh Sairam

Abstract In mobile ad hoc networks, routes are mostly multi-hop and nodes communicate via packet radios. With increasing number of mobile devices, the network is subjected to frequent topology changes. In such networks, a major challenge is to provide seamless connectivity without incurring excessive routing overhead. Routing protocols for ad hoc networks generally select the best possible path based on the current topology. However, these protocols cannot handle mobility induced route failures. A few ad hoc routing protocols have been proposed which select multiple, disjoint path to the destination. These multipath protocols start forwarding packet on the first available route and in case the route fails, it switches over to the next available path. In this way, they save on the route rediscovery overhead, but there is still a substantial overhead in switching among the alternate paths. In this paper, we propose to use link availability estimation technique to select the most stable route from among the alternate paths. The proposed route selection technique was tested by extending AOMDV, a well-known ad hoc, on-demand, multipath routing protocol. Empirical results indicate that selecting stable route results in higher throughput in dynamic network topology scenarios.

Keywords Mobile ad hoc network • Mobility prediction • Stable routing Multipath routing

A. P. Mazumdar (✉) · A. Surange · A. S. Sairam
Department of Computer Science and Engineering, Indian Institute of Technology Patna, Patna, Bihar, India
e-mail: arka@iitp.ac.in

A. Surange
e-mail: adhar@iitp.ac.in

A. S. Sairam
e-mail: ashok@iitp.ac.in

R. Maringanti et al. (eds.), *Proceedings of Ninth International Conference on Wireless Communication and Sensor Networks*, Lecture Notes in Electrical Engineering 299,
DOI: 10.1007/978-81-322-1823-4_5,

5.1 Introduction

A wireless mobile ad hoc network (MANET) is characterized by the lack of a central infrastructure and mobile nodes. The nodes act as hosts as well as routers and are assumed to be mobile. With the advent of new devices, the nodes are no longer restricted to laptops, but are getting more and more ubiquitous like PDAs and tablets. Due to the dynamic sets of users, user mobility, and varying channel conditions the network is subjected to frequent topology changes. This presents a challenging issue for design of routing protocols. Frequent changes in the topology can slow down communication or cause packet loss. Moreover, it incurs a huge overhead due to the need to re-establish new routes.

In this work, we present a routing protocol that adapts to frequent network changes. Unlike traditional protocols which select the best path, the idea here is to select the most stable route. The source uses ad hoc on-demand multipath distance vector (AOMDV) [1] protocol to find multiple, disjoint routes to the destination. AOMDV starts data transmission with the best route. In case the route fails, it switches over to an alternate path. The advantage of AOMDV is that it saves on the communication overhead to establish new routes to the same old destination. However, in the case of node mobility induced route failure, there is no guarantee that the alternate path will succeed.

Mobility prediction is a technique to predict future changes in the network topology emerging from the node's mobility. From among the multiple, disjoint paths, we use mobility prediction techniques to find the most stable route. This ensures that the chosen route will work best when nodes are mobile although the chosen route is not necessarily the shortest path. Experimental results show that stable routes give higher throughput in the event of low to high node mobility.

The organization of the paper is as follows:

In Sect. 5.2 we discuss the background of our work: existing multipath routing protocols and mobility prediction models. Section 5.3 describes our proposed technique for selecting stable routes. Experimental results and analysis is given in Sect. 5.4. Finally the concluding remarks are given in Sect. 5.5.

5.2 Background

In this section, we discuss several ingredients of mobile ad hoc networks as presented and discussed in the literature. As our goal is to modify AOMDV, we first discuss the principles of the protocol. Next, we present a short discussion on the link estimation techniques for wireless channels.

5.2.1 Single Versus Multiple Path

The well-known AODV [2] protocol was proposed to find the best path to the destination as and when required. In this protocol, whenever a node requires to route traffic to a destination node, it initiates a route determination process by broadcasting route request (RREQ) packet for the node. Intermediate nodes forward (broadcasts) message toward destination. When a node re-broadcasts a RREQ, it sets up a next-hop entry for the source of the packet in its routing table. Broadcasting RREQ ensures that the request packet reaches the destination through the shortest path. The destination unicasts a route reply (RREP) packet to the source and subsequent RREQ are discarded. The RREQ is sent along the reverse path created by the intermediate nodes. The next-hop entries created by the intermediate nodes while forwarding the packet is populated with information from the RREP. In this way routing table entries are setup and the path is established.

AOMDV [1] extends the AODV protocol to discover several paths between a traffic source and destination in every route discovery. Since the RREQ packet is flooded, nodes may receive several copies of the same RREQ having travelled through different paths. This approach is an extension of AODV where only the first request packet is used in the route construction and the duplicate RREQs are discarded. However, in AOMDV the duplicate RREQs are examined and three alternate forward paths are constructed between the source and destination. AOMDV like AODV starts transmitting data to the destination through the first (best) path. In case the source is unable to forward data through the current path, the next alternate path is tried. This makes the protocol robust to mobility-related route failures. AOMDV stores the best possible path as well as two alternate paths for every destination. However, only those routes which are one hop longer than the shortest path are chosen as possible candidate for alternate paths. The advantage of AOMDV relative to AODV, in such scenarios, is in terms of robustness to mobility-related route failures that result in higher throughput as the route discovery frequency is substantially reduced.

The alternate paths chosen by AOMDV, however, are not guaranteed to be functional or more stable in the presence of mobility-induced route failures. AOMDV keeps on trying alternate paths in the hope that at least one of them will be *alive*. In scenarios, where alternate paths also fail, AOMDV will perform worse than AODV. In a network where nodes are known to be mobile, a better route selection strategy would be to select the most stable one among the alternate routes available.

5.2.2 Mobility Tracking in Ad Hoc Networks

A model for tracking the mobility of users in MANET was proposed by Zainab et al. [3]. A dynamic linear system is used to model the motion of a node. The system is driven by a discrete command process. In order to predict the state of a node, the model considers the node's current position, velocity, and acceleration. The prediction process uses a modified version of *Kalman* filter along with a hidden semi-Markov model. Although their model can predict the future state of nodes, a major drawback is that it requires relative positions of the nodes. Moreover, the computational complexity is also high.

Su et al. [4] proposed using GPS position information to estimate the link expiration time between two adjacent mobiles. The model proposes to use this prediction to rediscover routes before they expire. Their goal is similar to that of this paper, in that they want to provide a seamless connectivity service by reacting before the connectivity breaks. However, the use of GPS in each node is a major overhead.

Capkun et al. [5] proposed a distributed algorithm that allows nodes within a network area to find their relative positions. The prediction is based on local information only. Their primary goal is to build a network coordinate system using range measurements between the nodes. Although this model do not use external information like GPS, their technique do not provide an estimate of time a link will be available between two nodes.

A link availability estimation technique is one which can predict the future state of wireless network given the current network state. In [6], estimate of the link availability is divided into two subproblems—first estimate the link availability when nodes are approaching and in the second case when they are receding. Transmit and received signal power is used to measure the instantaneous distance of two nodes. This information is then used to predict the time the link will expire. In this paper, we use this approach to find the most stable link. Details are explained in the following section.

Although, multipath routing and change in route prior to path expiration have been well studied, the benefits of both schemes have not been exploited for mobile network scenarios.

5.3 Stable Route Selection

In this section, we present our proposed methodology for selection of stable routes. The proposed stable route selection strategy is twofold. The first phase is route discovery, where multiple routes are found between the source and the destination using the steps of AOMDV. The second phase is the mobility prediction phase. For each route found, link availability estimation strategy is used to compute the stability of the route. The protocol then uses the most stable route to forward traffic.

5.3.1 Route Discovery

The route discovery process is same as that of AOMDV. A source node that needs to route traffic to a destination, initiates a route discovery process by broadcasting RREQ packets. As the route request packets are flooded, nodes generally receive multiple copies of the same RREQ. Like in AOMDV, these duplicate copies are used to our advantage to form alternate routes. Each duplicate packet is examined for potential alternate routes. However, while constructing the alternate routes

Destination	Sequence number	Advertised hop counts	Route list				
			Next Hop 1	Last Hop1	Hop count1	Timeout 1	LAE1
			Next Hop 2	Last Hop2	Hop coun2	Timeout 2	LAE2
			.	.	.	.	.
			.	.	.	.	.
			.	.	.	.	.

Fig. 5.1 Routing table

between the source and destination nodes, care has to be taken that the paths are disjoint and loop free.

In order to construct the alternate paths, when a forwarding node obtains a duplicate RREQ, it checks if more than one forwarder path exist to the destination. In case there is more than one path to the destination, a RREP is generated and send to the source along the alternate path. The reply packet includes a forward path to the destination that was not included in any of the previous RREPs. In such cases, the RREQ is not propagated any further. Otherwise, the intermediate node checks if the RREQ copy resulted in the construction of a reverse path. If not, the RREQ packet is further re-broadcasted.

The route reply packets generated by the destination is exactly same as that of the intermediate nodes. A RREP packet is generated by the victim for every copy of RREQ packet received, provided the request packet arrives via a loop free and disjoint alternate path.

In order to route packets along the reverse path, the last hop followed by the packets must be maintained in the routing table. This requires that the request and reply packets must also include the node address of the last hop. Practically, the last hop of the forward path will be first hop of packets routed in the reverse path.

5.3.2 *Routing Table*

Figure 5.1 shows the routing table used in our modified AOMDV protocol. The table has a new entry *LAE* which accords to *link availability estimation* of that determined path. Moreover, the proposed protocol being a diversified one, each list is required to store diverse path between node and destination. For every RREP packet received by a node, it calculates link availability estimation for that particular RREP and maintains its value in *LAE* field of the routing table.

5.3.2.1 Link Availability Estimation

Consider two nodes x and y, let $\{x, y\}$ be the link between them. Link Availability Estimation $L(T)$ is defined as the probability that the link will remain alive at time $T + t_0$ given it is available at t_0.

$$L(T) = P_r\{\text{link } \{x, y\} \text{ last from time } t_0 \text{ to } T + t_0\} \tag{5.1}$$

Given T, the link expiration time is an estimate of the link that is continuously available time between the two nodes. Let v be the relative velocity between the two nodes, d the instantaneous distance between the nodes, and R a normally distributed random variable. There can be three cases (1) nodes are approaching toward each other (2) nodes are moving away, and (3) nodes are stationary. The value of link expiration time for these three cases can be computed by the following formulas [7]:

- For approaching nodes:

$$T = \frac{1}{2v}\left(\sqrt{2d^2 - 4(d^2 - R^2)} + d\sqrt{2}\right) \tag{5.2}$$

  - For receding nodes:

$$T = \frac{1}{2v}\left(\sqrt{2d^2 - 4(d^2 - R^2)} - d\sqrt{2}\right) \tag{5.3}$$

  - For relatively stationary nodes T is infinite

The calculation of $L(T)$ can be divided into two parts: $L_1(T)$ in which the nodes movement remains unchanged, i.e., they continue to move in the same direction and the same speed and $L_2(T)$ for all other cases. That is,

$$L(T) = L_1(T) + L_2(T) \tag{5.4}$$

Assuming mobility epoch (a random length interval during which node movement is unchanged) is exponentially distributed with mean λ^{-1} the value of $L_1(T)$ is given by the following expression [6]:

$$L_1(T) = e^{-2\lambda T} \tag{5.5}$$

The calculation of $L_2(T)$ is difficult due to difficulties in learning link status between two nodes caused by nodes movement. Jiang et al. [6] proposed to find the approximate value $L_{\min}(T)$ of $L_2(T)$ as follows:

$$L_{\min}(T) = \frac{(1 - e^{-2\lambda T})}{2\lambda T} + \frac{\lambda T e^{-2\lambda T}}{2} \tag{5.6}$$

Once a node receives an RREP message it calculates the approximate distance d of the node from which the message was received using the received signal strength as follows:

$$d = \sqrt{\frac{P_t G_t G_r A_t^2 A_r^2}{P_r l}} \tag{5.7}$$

Here P_t and P_r are transmitted and received power, A_r and A_t are altitude of receiver and transmitter antenna, G_r and G_t are gain of receiver and transmitter antenna, and l is the path loss. It is assumed that nodes while transmitting message include the transmitted power, antenna altitude, and gain.

Using the value of d as computed above, the value of T can be computed using Eqs. 5.2 and 5.3. Finally, substituting the value of T into Eqs. 5.4, 5.5, and 5.6, link availability estimation between nodes can be computed as

$$e^{-2\lambda T} + \frac{(1 - e^{-2\lambda T})}{2\lambda T} + \frac{\lambda T e^{-2\lambda T}}{2} \tag{5.8}$$

Our protocol being an extension of AOMDV, we compute three routes to the destination and their $L(T)$. From these, we choose the route with maximum $L(T)$ to forward packet, since this is the path with the highest probability that it will be alive till time T.

5.4 Results and Analysis

To study the performance of our proposed protocol, necessary extensions were made to AOMDV. Our aim is to evaluate the effectiveness of our protocol when nodes are mobile. The mobility model used is the random way point model [8]. In this model, a node movement is described using pause time and motion period. A node stays at a particular position for a time called *pause time* before moving to the next position in some randomly selected direction with speed ranging between [0, Maxspeed].

The entire simulation was carried out in ns-2.35 [9]. This tool is used as it is suitable for designing new protocols, as well as comparing different protocols and traffic evaluations. The second important reason for choosing ns-2 is that AOMDV and its predecessor AODV have been implemented and tested in this tool. Thus, extending AOMDV to support stable routes become convenient. The number of nodes considered was 100 spread over an area of 500×500 m^2. The traffic model considered was CBR/VBR as background traffic and ftp traffic between the source and destination pair. Each simulation run was for about 500 s with an initial warm-up period of 200 s.

The following performance metrics have been considered in this paper:

1. **Packets Dropped**: Percentage of packets dropped during data transfer. The packet loss may be due to error in the channel or non-availability of next forwarder.
2. **Route Flaps**: This metric will give a measure of the number of route changes. Higher route flaps will cause greater number of packets dropped.
3. **Throughput**: This metric measures the average number of packets successfully delivered. The metric will give a measure of a protocol's speed in forwarding packets.

In this work, we compare our protocol with AOMDV. Route discovery overhead has not been used as a metric, since it is the same for both the protocols. To analyze the effect of mobility, pause time was varied from 10 s (high mobility) to 100 s (low mobility). The speed of the nodes vary from 0 to 20 m/sec.

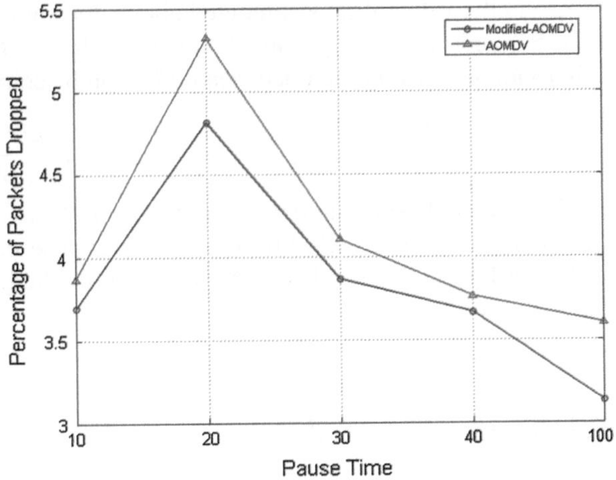

Fig. 5.2 Comparison of packet loss for different pause times

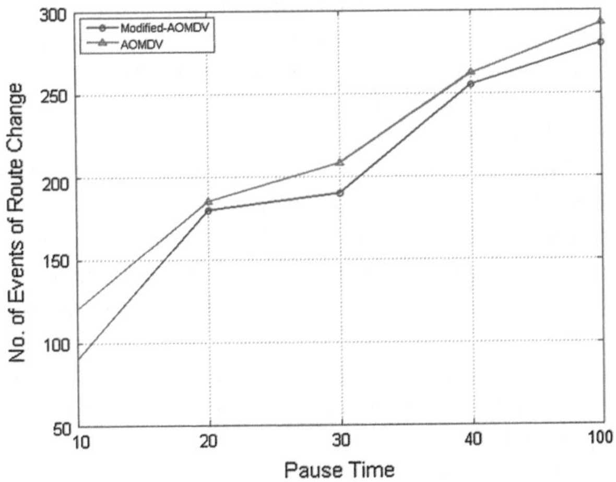

Fig. 5.3 Comparison of route flap for different pause times

Figure 5.2 compares the percentage of packets dropped between AOMDV and our protocol for different pause times. As can be seen from the figure, the packets dropped are always less in the case of our proposed protocol.

To confirm that the higher percentage of packets dropped in case of AOMDV is because of route flap, we compare the percentage of route flaps for both the cases. From Fig. 5.3 it can be seen that the number of route changes is higher in the case of AOMDV, which accounts for the higher packet loss. AOMDV changes route when the current is no longer able to forward packets. We observe that packets are dropped during these route flaps. AOMDV selects path based on first come first serve basis, whereas our protocol selects the most stable path. Hence, route flap is significantly reduced.

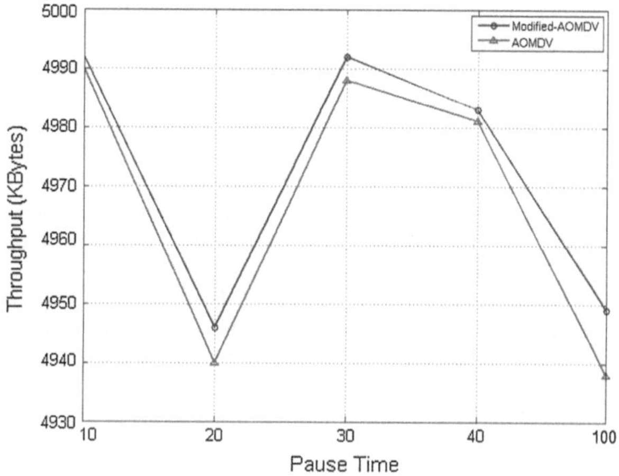

Fig. 5.4 Comparison of throughput for different pause times

Finally, in Fig. 5.4 the throughputs of both the protocols is compared. Our protocol has a higher throughput than AOMDV for all the cases. This result confirms our hypothesis that selecting a route which will be available during the period of entire data transfer is better than selecting the best route and changing to an alternate route later.

5.5 Conclusion

In this work, we proposed a multipath routing protocol for MANET that selects the most stable route instead of the first available path. Link availability estimation technique was used to choose the most stable route among the multiple paths available. The protocol was implemented by extending AOMDV, an existing multipath routing protocol for ad hoc networks. AOMDV can deal with mobility-induced route failures as opposed to its single path counterpart AODV. However, during route failures, the overhead to change from one route to an alternate path is high. The proposed protocol was tested in a network where node speeds ranged from low to very high. In all the scenarios, our proposed extension to AOMDV reduces packet loss, decreases route flaps, and as a result improves throughput.

References

1. Marina, M.K., Das, S.R.: Ad hoc on-demand multipath distance vector routing. Wirel. Commun. Mob. Comput. **6**(7), 969–988 (2006)
2. Perkins, C.E., Royer, E. M.: Ad hoc on-demand distance vector routing, In: 2nd IEEE Workshop Mobile Computing System and Applications pp. 90–100, Feb 1999
3. Zaidi, Z.R., Mark, B.L.: A mobility tracking model for wireless ad hoc networks. Wirel. Commun. Netw. **3**, 1790–1795 (2003)

4. Su, W., Lee, S-J., Gerla, M.: Mobility prediction and routing in ad hoc wireless networks. MOBICOM (1998)
5. Capkun, S., Hamdi, M., Hubaux, J-P.: GPS free positioning in mobile ad hoc networks. Cluster Computing, Kluwer Academic Publishers, **5**(2), April (2002) ISSN: 1386-7857
6. Jiang, S., He, D., Rao, J.: A prediction-based link availability estimation for mobile ad hoc networks, INFOCOM 2001. In: Twentieth Annual Joint Conference of the IEEE Computer and Communications Societies, vol. 3, pp. 1745–1752
7. He, D.J., Jiang, S.M., Rao J.Q.: Link availability prediction model for wireless ad hoc networks, In: Proceeding 2000 International Conference on Distributed Computing System Workshop, Taipei, Taiwan, pp: D7–D11, Apr (2000)
8. Broch, J., Maltz, D.A., Johnson, D.B., Hu, Y.C., Jetcheva, J.: A performance comparison of multi-hop wireless ad hoc network routing protocols. In: Proceedings of the 4th annual ACM/IEEE international conference on mobile computing and networking, MobiCom '98, pp: 85–97, ACM (1998)
9. NS-2, *Network Simulator version 2*, http://nsnam.isi.edu/nsnam/index.php/Main_Page

Chapter 6
p-shrink: A Heuristic for Improving Minimum All-to-All Power Broadcast Trees in Wireless Networks

Wilson Naik Bhukya and Alok Singh

Abstract All-to-all broadcast refers to a network scheme where every node communicates to every other node in the network; all-to-all broadcast problem uses minimum unique cast tree (MUCT) scheme to generate power efficient all-to-all broadcast tree (BT) in wireless networks. The minimum all-to-all power broadcast problem is shown as NP-complete. In this paper, we have presented the *p-shrink* procedure, a heuristic to find all-to-all minimum power broadcast trees in wireless networks. Specifically, we focus on low complexity Kruskal's minimum spanning tree algorithm and show through extensive experimentations that power-efficient broadcast trees are obtained almost always, with considerably minimum all-to-all power with the proposed heuristic. The simulation results on numerous bench mark problem instances with 20, 50, and 100 vertices confirm that the proposed heuristic significantly reduces the total all-to-all power cost of broadcast trees .

Keywords MUCT • Broadcast tree • MEB • All-to-all broadcast • Anycast

6.1 Introduction

Wireless networks in different forms as ad hoc networks and sensor networks have received significant attention in recent years due to their use in critical applications from civil to military domains. Unlike wired networks or cellular network, a wireless node has no fixed backbone support. Usually, each node in these networks is not

W. N. Bhukya (✉) · A. Singh
School of Computer and Information Sciences,
University of Hyderabad, Hyderabad 500046, India
e-mail: naikcs@uohyd.ernet.in

A. Singh
e-mail: alokcs@uohyd.ernet.in

R. Maringanti et al. (eds.), *Proceedings of Ninth International Conference on Wireless Communication and Sensor Networks*, Lecture Notes in Electrical Engineering 299,
DOI: 10.1007/978-81-322-1823-4_6, © Springer India 2014

equipped for frequent recharging or replacement of battery power. This highlights the importance of power efficient algorithms to maximize the broadcast or multicast communication and also to increase the overall lifetime of the network. Broadcasting or multicasting of one-to-all or all-to-all communication is one of the fundamental operations in any form of communication including wireless. Power efficient heuristics are need of the hour to optimize broadcast and multicast schemes in wireless networks.

Wireless networks have a broadcast advantage that are fundamentally different from wired networks. When a sender uses an omnidirectional antenna, every data transmission by the sender can be received by all nodes within its transmission range without any additional cost to the sender. This property is known as wireless multicast advantage (WMA) [1].

Broadcast is a mechanism (one-to-all) in which message is sent by an identified source node to the rest of the nodes in a particular network. In multicast, the communication messages are sent only to the subset of nodes. Convergcast (all-to-one) is a dual of broadcast since the data flows back to the sender. Thus the total transmission cost spent in convergcast, which uses the same transmission paths, as a tree rooted at the sink node, is the sum of the energy spent by each node to transmit that packet to its parent. In anycast (All-to-All) broadcast, every node is a root for broadcasting and constructing *n* individual broadcast trees for an *n-node* network is unworkable due to high power requirement for transmission and computation. Additionally, it poses risk of BT recomputation even for a minor topology change. One of the basic solutions, although most expensive in terms of power, is to permit all the nodes in the network to communicate to every other node directly.

For a given network with a given source node, the minimum power broadcast (MPB) problem in wireless networks is to broadcast to rest of the nodes, either by direct transmission or indirectly through hoping, such that the overall transmission power is minimized. It is shown that MPB problem in wireless networks is NP-complete, implying that optimal polynomial time algorithms are unlikely to exist [2, 3]. In this paper, we present *p-shrink,* a heuristic for improving minimum power all-to-all broadcast trees in wireless networks.

The rest of the paper is organized as follows: Sect. 6.2 presents related work. In Sect. 6.3, we present construction of all-to-all broadcast trees with BIP(Broadcast Incremental Power), Kruskal's, and proposed heuristic *p-shrink*. Experimental results and discussion are presented in Sect. 6.4. Section 6.5 outlines some concluding remarks.

6.2 Related Work

The minimum power broadcast problem in wireless networks has gained much attention over the last few years. Much focus of research is mainly on the one-to-many communication pattern. In the literature, the minimum power broadcast problem is converted as finding a minimum spanning tree (MST); many algorithms are devised with these criteria of finding MST. Minimum Energy Broadcast Trees (MEBT) and Minimum Energy Multicast Trees (MEMT) [1–5]. Wieselthier et al. [1] proposed Broadcast Incremental

Power Algorithm (BIP) which was node based rather than a traditional link-based approach. It was a modified version of the Prim's algorithm to construct MST. During each step of BIP, it added an uncovered node to the tree with minimum cost. However, little attention has been paid to all-to-all broadcasting and multicasting that is similar to many-to-many communication type in the literature. The functional applications of all-to-all broadcasting can be seen in many scenarios mainly in distributed gaming; ad hoc classrooms; convention centers; alerting systems, etc.

All-to-all broadcast is very expensive in terms of power. Very little work has been done for this type of communication. Existing research results for single [all-to-all] or source-dependent broadcast trees [1, 6] are not efficient for all-to-all broadcast problem, since they are not designed for back communication to the broadcast initiator. It is unfeasible and complex to generate an exclusive BT for each terminal node for *n-node* networks. As a result, the total all-to-all broadcast power is the sum of the power consumption of these broadcast trees.

A more realistic approach to this problem is to build one BT shared by all. In this approach, energy and delay overheads are negligible. Further, this approach makes shared BT maintenance simpler, because the power transmission regulation of each node is fixed to one broadcast tree. Boien et al. [7] proposed a maximum unique cast graph (MUC) problem. Given a wireless network (V, w), assigning enough power levels to all the nodes in the network such that the unidirectional communication links among the nodes make a strongly connected graph that can be used for all-to-all broadcast. Further [7] proposed an approximation scheme named minimum unique cast tree (MUCT) to generate a shared BT using BIP algorithm to minimize the all-to-all broadcast power. They computed and compared the results with different approximation algorithms BIP, IMBM, and WMA-MST [1, 8] with MUCT. Naik and Singh [9] have implemented all-to-all broadcast with known approximation algorithms and have shown that Kruskal's way of generating all-to-all shared BT works better than other approaches reported. Naik and Singh [10] have proposed an energy efficient algorithm for this problem which is optimal than the other approximation algorithms considers internal nodes for decision making for computing best broadcast trees. In this paper, we proposed a heuristic *p-shrink* to improve the total all-to-all shared broadcast trees power which emphasizes the need of leaf nodes for construction of broadcast trees, simulation results show that our *p-shrink* performs better in terms of minimum all-to-all power than other approaches including the results reported by us [10].

6.3 Construction of Shared Broadcast Trees for Minimizing All-to-All Communication Power

6.3.1 Algorithm (MUCT) Minimum Unique Cast Tree

New measure proposed by [7] to measure the power used by node i for the graph $G = (V, E)$, using following equations.

$$\text{power}_i^T = \max_{j \in si} \quad \text{cost}_{ij} \tag{6.1}$$

Energy at a node based on distance to the farthest sibling. power_i^T the total power of all the nodes in the tree T.

$$T : \text{power}^T = \sum_{i \in v} \text{power}_i^T \tag{6.2}$$

Algorithm 1 *MUCT: Minimum Unique Cast Tree*

Let A =BIP, r-shrink, Kruskal's
Input is Read: (V, w) ;
Initialization:Let $power_0 = \infty$and $OPT = \phi$
Main Procedure: For all r ∈ V do
Build a broadcast tree BT rooted at r, using an approximation algorithms A.
Let T to be un oriented tree obtained by ignoring the orientation in BT.
Define power level of all the nodes as equation(1)
Compute the $power^T$ as in equation(2)
if ($power^T < power_0$) then set OPT to T and $power_0 = power^T$ End for

6.3.2 MUCT with Kruskal's Algorithm for Generating Shared Broadcast Tree (BT)

Since *r-shrink* heuristic [9, 11] was not power efficient on BIP which generates MST in Prim's way, we have explored this problem with Kruskal's algorithm [12] for generating MST. Kruskal's algorithm showed good results compared to BIP and BIP with *r-shrink*. The total all-to-all power value of shared BT with 20 node bench mark instance using Kruskal's algorithm = 1,153,819 where as BIP = 1,181,146.710 and BIP with *r-shrink* = 2,026,171.370.

6.3.3 MUCT with Power Efficient Algorithm for Generating Shared Broadcast Tree (BT) [10]

We have generated a shared BT for minimizing all-to-all communication using power efficient algorithm proposed by [10] as approximation algorithm and used MUCT [7] algorithm for computing the total power. The total all-to-all power value of shared BT with 20 node bench mark instance using power efficient algorithm = 1128387.220 which is more optimal compared to Kruskal's. The drawback of power efficient algorithm for all-to-all is its limitation by excluding all the leaf nodes which plays crucial role in minimizing the overall power for minimum power broadcast problem for all-to-all.

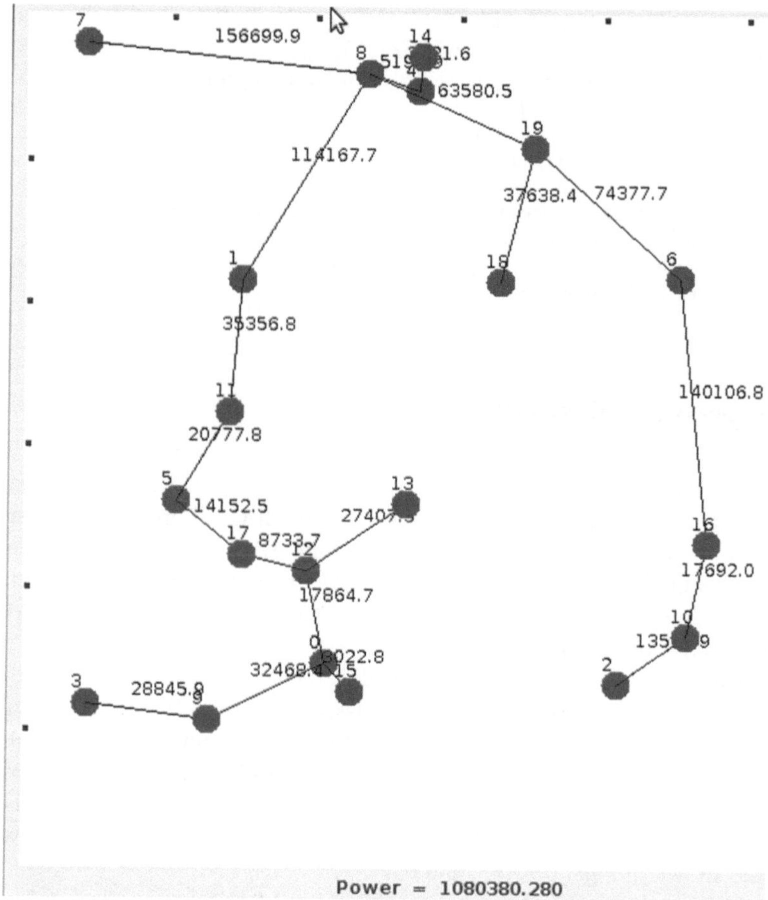

Fig. 6.1 *p-shrink* heuristic based shared broadcast tree for 20 node network with bench mark problem instance for minimizing all-to-all broadcast power

6.3.4 MUCT with p-shrink Heuristic for Generating Shared Broadcast Tree (BT)

In *p-shrink* heuristic, we have considered all the vertices including leaf nodes compared to [10] where they have considered only the internal nodes of the BT similar to [11]. In this approach, we have realized that leaf nodes play a crucial role in minimizing the broadcast power which was not the case in other algorithms [10, 11]. The total all-to-all power value of shared BT for 20 node bench mark instance using Kruskal's algorithm = 1,153,819.830 whereas BIP = 1,181,146.710 and BIP with *r-shrink* = 2,026,171.370. Energy efficient algorithm = 1,128,387.220, and *p-shrink* (considering leaf nodes) = 1,080,380.280 (Fig. 6.1).

Algorithm 2 ***p-shrink: Heuristic***

Build a broadcast tree BT using Kruskal's MST algorithm.
For each v$\in$ **G.V:** Traverse all the vertices of BT including leaf nodes and find vertices whose Degree >1
Find maximum associated edge of BT.
Delink BT into subtrees BT1, BT2 using maximum associated edge of the BT.
Find possible minimum associated edge cost from subtrees BT1 and BT2 to make minimal all-to-all cost tree .
Join the BT1 and BT2 using minimum associated edge cost such that overall power of tree BT is MINIMAL

6.4 Experimental Results

In order to validate the effectiveness of our work, we have conducted extensive experiments to compare total minimum power for all-to-all broadcast power based on shared broadcast trees generated by algorithms BIP, BIP + *r-shrink*, Kruskal's, power efficient algorithm and proposed *p-shrink* heuristics. We have implemented these algorithms in Java and executed on Intel Core 2 Quad with 4 GB RAM running at 2.8 GHz under Linux environment.

Experimental results conducted on different networks as well on bench mark data using BIP, BIP + *r-shrink*, Kruskal's, energy efficient algorithm, and *p-shrink*. *p-shrink* heuristic best suited for constructing minimum all-to-all broadcast trees compared to all the previous approaches almost always. We have also simulated our approach on benchmark dataset problem instances used for minimum energy broadcast problem (MEB) on 20,50, and 100 vertices with 30 instances shown in Figs. 6.2, 6.3, and 6.4.

For each network instance *k*, let *Pmuc-bip, Pmuc-bip* + *r-shrink, Pmuc-Kruskal, Pmuc-eeA,and Pmuc-p-shrink* be the minimum total energy used for all-to-all broadcast using the shared broadcast trees generated by BIP, BIP + *r-shrink,* Kruskal's, Energy efficient, and *p-shrink*, respectively. We select the minimum energy tree among the trees generated by the above mentioned algorithms. *Pmin = minimum of* {*Pmuc-bip, Pmuc-bip + r-shrink, Pmuc-eeA, Pmuc-Kruskalway Pmuc-p-shrink*}.

Then we normalize *Pmuc-bip, Pmuc-bip + r-shrink, Pmuc-eeA, Pmuc-Kruskalway Pmuc-p-shrink*

Pmin: = {Pmuc-bip = {Pmuc-bip/Pmin}, Pmuc-bip + *r-shrink* = {Pmuc-bip + *r-shrink*/Pmin}, Pmuc-Kruskalway = {Pmuc-Kruskalway/Pmin}, Pmuc-eeA = {Pmuc-eeA/Pmin}, Pmuc-*p-shrink* = {Pmuc-*p-shrink*/Pmin}}. The normalized energy values associated with the broadcast shared trees generated by an algorithm is independent on the size of distance scaling factor.

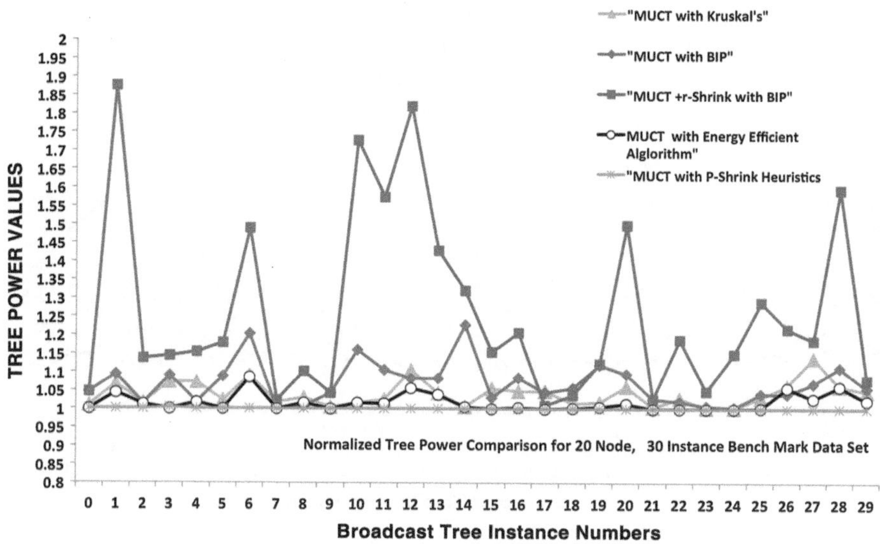

Fig. 6.2 Normalized total tree power results for different shared broadcast trees with bench mark data with 20 vertices and 30 instances

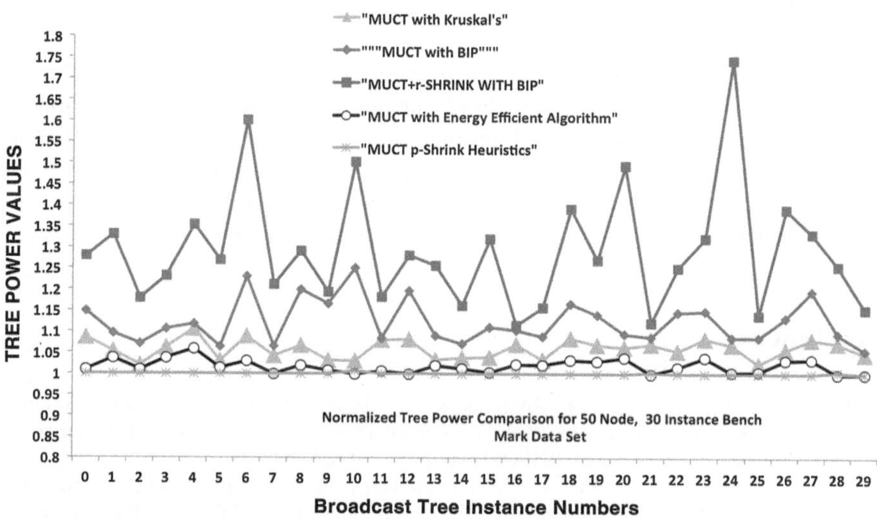

Fig. 6.3 Normalized total tree power results for different shared broadcast trees with bench mark data with 50 vertices and 30 instances

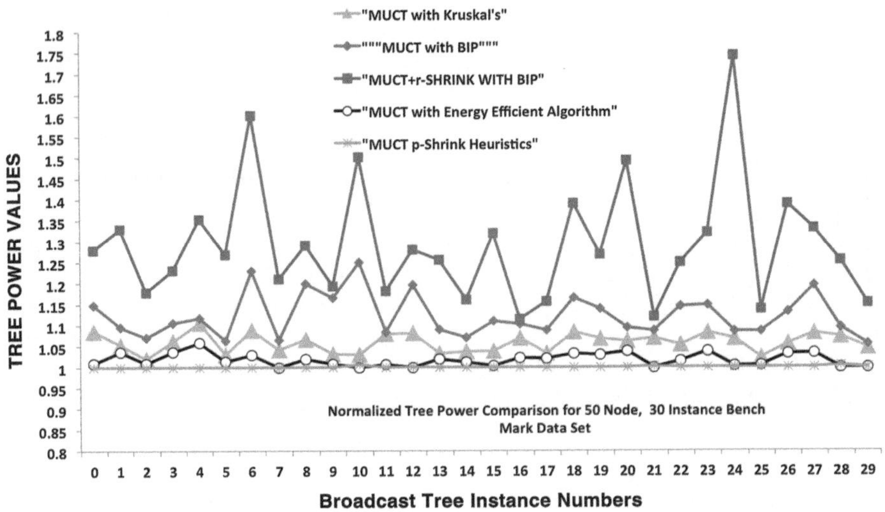

Fig. 6.4 Normalized total tree power results for different shared broadcast trees with bench mark data with 100 vertices and 30 instances

6.5 Conclusions

In this paper, we have proposed *p-shrink* heuristic for improving total all-to-all power in broadcast trees and compared our approach with other approximation algorithms reported on bench mark problem instances with 20, 50, and 100 nodes with 30 instances. Experimental results validate the effectiveness and power efficiency in minimizing total all-to-all power broadcast trees. As a future work, we plan to further optimize our algorithm in terms of total all-to-all power.

References

1. Wieselthier, E., Nguyen, G.D., Ephremides, A.: On the construction of energy- efficient broadcast and multicast trees in wireless networks. In: Proceedings of The IEEE Conference on Computer Communications (INFOCOM), p. 58559 (2000)
2. Cagalj, M., Hubaux, J., Enz, C.: Minimum-energy broadcast in all-wireless networks: NP-completeness and distribution issues. In: Proceedings of the Mobicom 2002 Conference, Atlanta, GA, 23–28 September 2002
3. Liang, W.: Constructing minimum-energy broadcast trees in wireless ad hoc networks. In: Proceedings of MOBIHOC??02, pp 112–122 (2002)
4. Singh, A., Naik, W.B.: A Hybrid genetic algorithm for minimum energy broadcast. J. Appl. Soft. Comput. **11**(1), 667–674 (2011)
5. Cagalj, M., Hubaux, J.-P., Enz, C.: Minimum-energy broadcast in all-wireless networks. Wirel. Netw. **11**, 177–188 (2005)
6. Papadimitriou, I., Georgiadis, L.: Minimum-energy broadcasting in multi-hop wireless networks using a single broadcast tree. Mob. Netw. Appl. **11**, 361375 (2006)

7. Bein, D., Zheng, S.Q.: Energy efficient all-to-all broadcast in wireless networks. J. Inform. Sci. **80**, 1781–1792 (2010)
8. Stojmenovic, I., Seddigh, M., Zunic, J.: Dominating sets and neighbor elimination- based broadcasting algorithm in wireless networks. IEEE Trans. Parallel Distrib. Syst. **13**, 1425 (2002)
9. Naik, W.B., Singh, A.: A study on energy issues in construction of all-to-all minimum power broadcast trees in wireless networks. In: Proceeding of, International Conference on Advances in Computing, Communications and Informatics (ICACCI-2013), Mysore, India
10. Naik, W.B., Singh, A.: Energy-Efficient algorithm for computing all-to-all minimum power broadcast trees in wireless networks. Paper has been communicated for review
11. Das A.K., Marks, R.J., Sharkawvi, M.E., Arabshahi, P., Gray, A.: r-shrink: A heuristic for improving minimum power broadcast trees in wireless networks. In: GLOBECOM-2003, pp. 523–528
12. Kruskal, J.B.: On the shortest spanning subtree of a graph and the traveling salesman problem. Proc. Am. Math. Soc. **7**(1), 4850 (1956)
13. Stojmenovic, I., Seddigh, M., Zunic, J.: Internal nodes based broadcasting in wireless networks. In: Proceedings of the Hawaii International Conference on System Sciences (HICSS), p. 10 (2001)
14. Chen, B., Jamieson, K., Balakrishnan, H., Morris, R.: SPAN: an energy-efficient coordination algorithm for topology maintenance in ad hoc wireless networks. ACM Wirel. Netw. J. **8**(5), 481494 (2002)

Chapter 7
Hook-Shaped Printed Multiband Antenna for Different Wireless and Mobile Applications

Amit Kumar Tripathi and B. K. Singh

Abstract In this paper, a printed monopole wireless antenna is presented which is a very simple and compact but very effective radiating element for the multiband wireless communication system and different mobile devices. This antenna simultaneously covers the following wireless bands: UMTS (1,920–2,170 MHz), 2.4-GHz WLAN (2,400–2,484 MHz), 5-GHz WLAN (5,150–5,350/5,725–5,825 MHz), ITS (5,795–6,400 MHz), Wi-Fi (5 GHz), MIMO applications and 3G band of 1.8–2 GHz. Proposed antenna is very compact and lightweight having area of 20 × 20 mm^2 and it is also cheap due to the use of FR4 substrate and less conducting metal. This antenna can be a good choice for the applications of WLAN and other wireless devices. This antenna has a very good efficiency and gain which can provide an effective radiation. Since antenna covers 1.8–2.3, and 4.2–9.2 GHz (61.11 %) so it is bandwidth efficient and having the efficiency of over 92 %. The antenna is designed and simulated using finite element method (FEM)-based simulator HFSS V13.0.

Keywords Key–monopole antenna • Microstrip • Parametric • Inverted f-antenna FEM

7.1 Introduction

In recent years, the mobile communication and wireless communication has grown exponentially. During this growth, the need of high data rate is a very big challenge for the communication engineers. The different applications of wireless

A. K. Tripathi (✉) · B. K. Singh
Departement of Electronics and Communication Engineering, BTKIT, Dwarahat, Uttarakhand, India
e-mail: amit_elex@hotmail.com

B. K. Singh
e-mail: bksapkec@yahoo.com

R. Maringanti et al. (eds.), *Proceedings of Ninth International Conference on Wireless Communication and Sensor Networks*, Lecture Notes in Electrical Engineering 299, DOI: 10.1007/978-81-322-1823-4_7, © Springer India 2014

communication have the requirement of such type of antenna which can be operated in different bands according to their applications. It is very important for the antenna designers that the size and weight of the antenna should be very small that can be embedded into the wireless handsets.

These antenna modules are expected to provide effective broad band matching, an acceptable gain, and consistent radiation pattern throughout the allotted frequency band.

Monopole antennas are considered best for the multiband applications due to some of their properties such as low cost, light weight, simple fabrication, and compact size, and so on [1–7].

Many multiband antennas are investigated such as inverted F-antennas, slot antennas, and planer antennas. For the application of the wireless local area networks (WLANS), worldwide interoperability for microwave and access (Wi-Max), and universal mobile communication system [6–12]. These antennas have the drawback of having complex analysis of operating bandwidth and larger size.

In this paper a monopole antenna is presented having very compact size and weight. It uses an arc-shaped single electrical path and provides resonance in wide band. This microstrip fed wideband antenna serves different wireless applications such as UMTS (1.920–2.170) GHz, Wi-Fi (5 GHz), WLAN (5.15–5.35/5.72–5.82) GHz, ITS, MIMO etc.

The change in the structure of an antenna from simple patch to an arc-shaped patch leads to conversion of antenna to wide-band. The size and cost of the antenna is also reduced due to less metallic structure available in reference with [13]. Here the reduction in the cost is in the reference with the losses that take place in the antenna; since this antenna structure has less radiating element, less ohmic losses will take place which may be compared with the cost effectiveness of the antenna. There is no need of tuning the antenna for different applications in wireless communication.

7.2 Antenna Geometry

The geometry of the proposed antenna is shown in the Fig. 7.1. It consists of an arc-shaped single electrical path which requires less metal as compared to [13]. The antenna is designed on the substrate FR4_epoxy having permittivity of $\varepsilon r = 4.4$, Tan $\delta = 0.024$, and thickness of 1.6 mm. The feed line is designed for the characteristic impedance of 50 Ω SMA connector available in the market with the corresponding width of Wf = 1.875 mm; the actual dimension of the antenna is $20 \times 20\ \text{mm}^2$.

Where $R1 = 3$ mm, $R2 = 4$ mm, Wp = 3.7 mm, Lf = 8 mm, Lp = 4.5 mm, and G-arm-R = 0.56 mm. The proposed structure is designed using the finite element method (FEM)-based simulator HFSS V13.0. The designed structure is shown in the Fig. 7.3. Figure 7.2 shows the ground structure of the proposed antenna. To get the optimized result, parametric analysis is performed with respect to the length '*L*' of

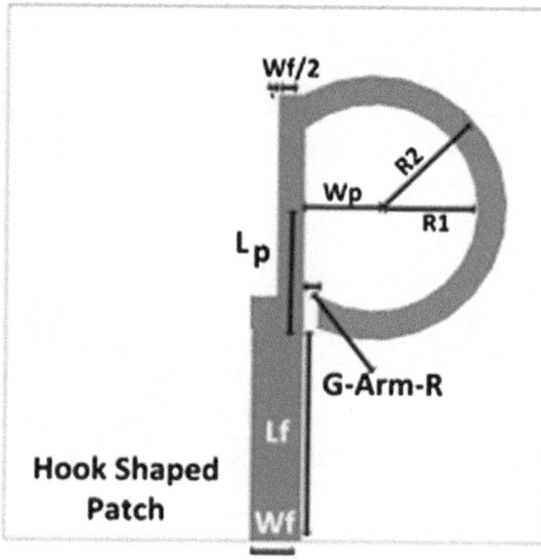

Fig. 7.1 Design structure of proposed antenna

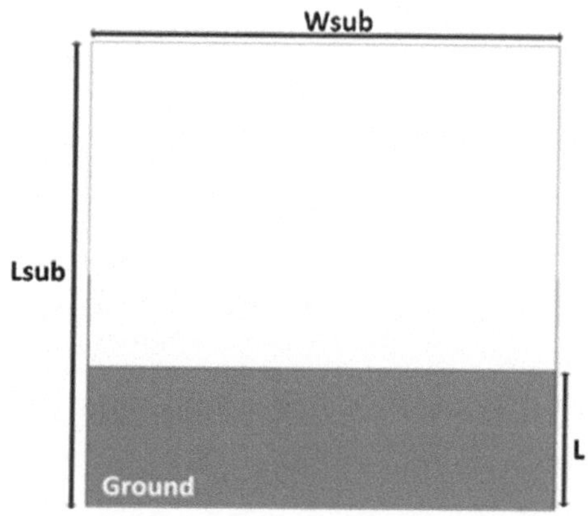

Fig. 7.2 Ground plane of the proposed antenna

the ground plane. L_{sub} denotes the length of the substrate, and have the value 20 mm, where as W_{sub} denotes the width of the substrate and have the same value as L_{sub}.

7.3 Simulation Results and Discussion

In this section, the simulation results of the designed antenna are presented. Through the optimization process the optimal results are found. Here the parametric analysis is used for the optimization of the result. Figure 7.4 shows the return loss S_{11} for the different values of the ground sheet length 'L'. From the variation of the L, we observed that extreme bandwidth is obtained at the value $L = 4$ mm from the origin along the Y-axis. The return loss S_{11} for the optimum value of 'L' is given in the Fig. 7.5. From Fig. 7.4, we can conclude that as we increase the

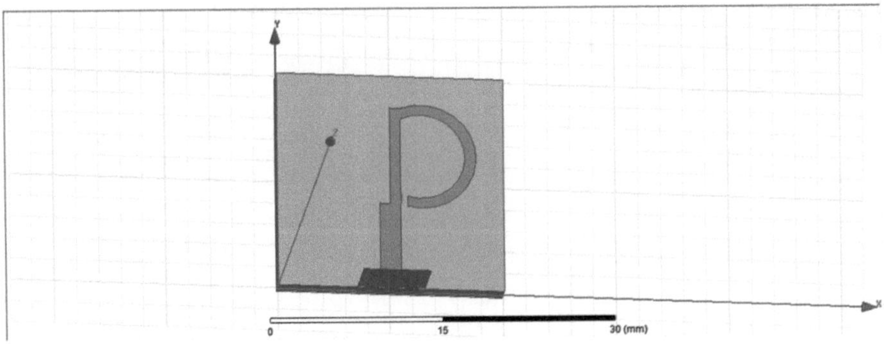

Fig. 7.3 Design of the proposed antenna using software HFSS V13.0

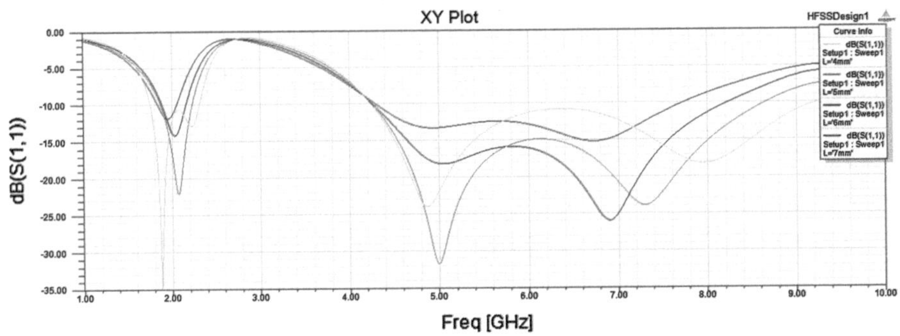

Fig. 7.4 Return loss S_{11} at different values of '*L*'

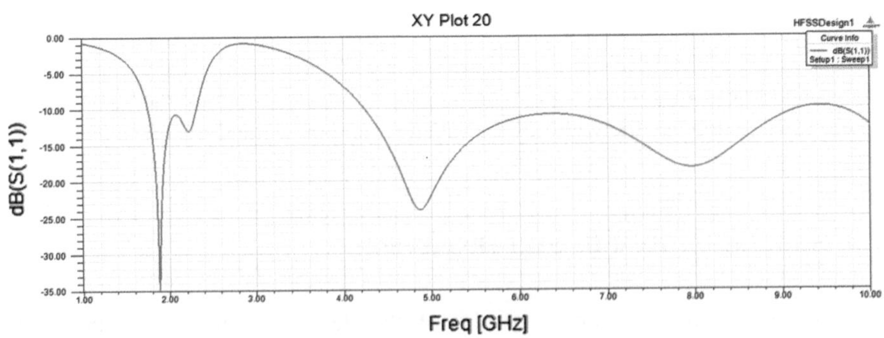

Fig. 7.5 Return loss S_{11} at $L = 4$ mm

value of '*L*' the bandwidth decreases. For the values lower than $L = 4$ mm, the band width increases again increases but that cannot be used in the proposed

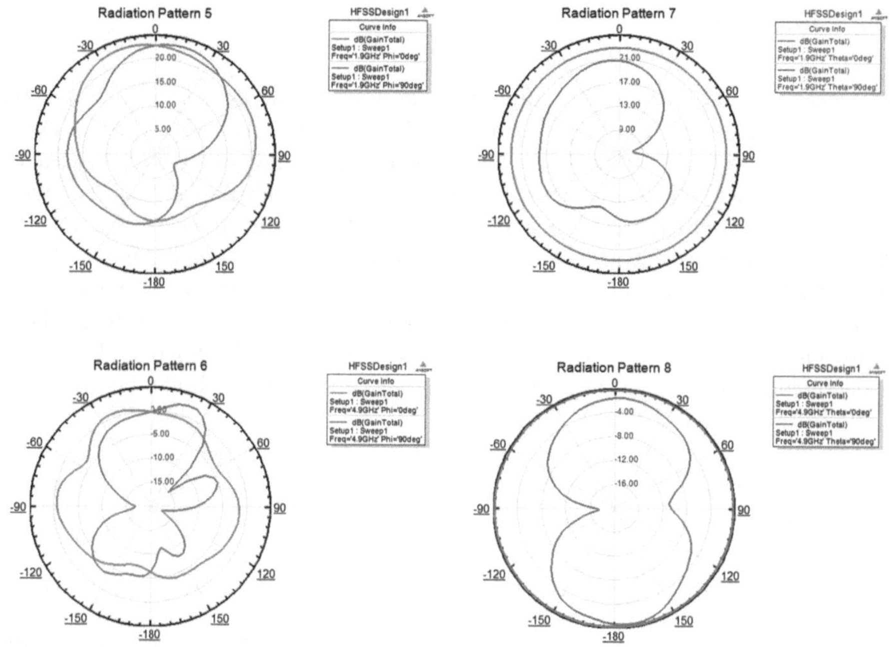

Fig. 7.6 Radiation patterns of proposed antenna at different frequencies. **a** Radiation pattern at 1.9 GHz, $\Phi = 0^{\circ}, 90^{\circ}$. **b** Radiation pattern at 4.9 GHz, $\Phi = 0^{\circ}, 90^{\circ}$. **c** Radiation pattern at 1.9 GHz, $\theta = 0^{\circ}, 90^{\circ}$. **d** Radiation pattern at 4.9 GHz, $\theta = 0^{\circ}, 90^{\circ}$

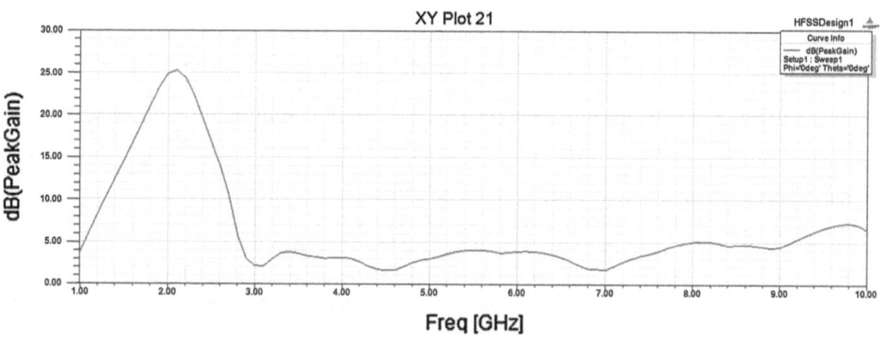

Fig. 7.7 Plot of frequency versus peak gain

applications so we have not included the values lower to $L = 4$ mm. Figure 7.6 shows the radiation pattern of the proposed antenna.

From Fig. 7.5 it can be observed that the resonating frequencies are 1.9 and 4.9 GHz so the whole analysis about radiation pattern gain etc. had been done on these resonating frequencies (Fig. 7.7).

7.4 Conclusion

This paper describes a novel configuration of a monopole microstrip-fed antenna that can operate in multiple band including UMTS, WLAN, ITS, MIMO, and Wi-Fi. Proposed antenna is very compact and lightweight having area of 20×20 mm^2 and it is also cheap due to the use of FR4 substrate and less conducting metal. This antenna can be good choice for the applications of WLAN and other wireless devices. This antenna has a very good efficiency and gain which can provide an effective radiation. There is no need of tuning because antenna covers all mentioned applications and having the return loss of below −10 dB in these application ranges. Since antenna covers 1.8–2.3 and 4.2–9.2 GHz (61.11 %) so it is bandwidth efficient and having the efficiency of over 92 %. It can be also used in 3G band of 1.8–2 GHz [14].

References

1. Liu, W.C., Wu, C.M., Dai, Y.: Design of triple-frequency microstrip-fed monopole antenna using defected ground structure. IEEE Trans. Antennas Propag. **59**(7), 2457–2463 (2011)
2. Qin, P.Y., Weily, A.R., Guo, Y.J., Bird, T.S., Liang, C.H.: Frequency reconfigurable quasi-Yagi folded dipole antenna. IEEE Trans. Antennas Propag. **58**(8), 2742–2747 (2010)
3. Latif, S.I., Shafai, L., Sharma, S.K.: Bandwidth enhancement and size reduction of microstrip slot antennas. IEEE Trans. Antennas Propag. **53**(3), 994–1003 (2005)
4. Liu, W.X., Yin, Y.Z., Xu, W.L.: Compact self-similar triple-band antenna for WLAN/WiMAX applications. Microw. Opt. Technol. Lett. **54**(4), 1084–1087 (2012)
5. Dastranj, A., Imani, A., Naser-Moghaddasi, M.: Printed wide-slot antenna for wideband applications. IEEE Trans. Antennas Propag. **56**(10), 3097–3102 (2008)
6. Naser-Moghadasi, M., Sadeghzadeh, R.A., Katouli, M., Virdee, B.S.: CPW-fed compact slot antenna for WLAN operation in a laptop computer. Microw. Opt. Technol. Lett. **52**(6), 870–873 (2010)
7. Pourahmadazar, J., Ghobadi, C., Nourinia, J., Shirzad, H.: Multiband ring fractal monopole antenna for mobile devices. IEEE Antennas Wirel. Propag. Lett. **9**, 863–866 (2010)
8. Cai, L.Y., Zeng, G., Yang, H.C., Cai, Y.Z.: Integrated bluetooth and multi-band ultra-wideband antenna. Electron. Lett. **46**(10), 688–689 (2011)
9. Lu, J.-H., Huang, B.-J.: Planar multi-band monopole antenna with L-shaped parasitic strip for WiMAX application. Electron. Lett. **47**(12), 671–672 (2010)
10. Pei, J., Wang, A., Gao, S., Leng, W.: Miniaturized triple-band antenna with a defected ground plane for WLAN/WiMAX applications. IEEE Antennas Wirel. Propag. Lett. **10**, 298–302 (2011)
11. Xu, P., Yan, Z.-H., Wang, C.: Multi-band modified fork-shaped monopole antenna with dual L-shaped parasitic plane. Electron. Lett. **46**(6), 364–365 (2011)
12. Lee, Y.-C., Sun, J.-S.: A new printed antenna for multiband wireless applications. IEEE Antennas Wirel. Propag. Lett. **8**, 402–405 (2009)
13. Naser-Moghadasi, M., Sadeghzadeh, R.A., Fakheri, M., Aribi, T., Sedghi, T., Virdee, B.S.: Miniature hook-shaped multiband antenna for mobile applications. IEEE Antennas Wirel. Propag. Lett. **11**, (2012)
14. Liang, J., Yang, H.Y.D.: Varactor loaded tunable printed pifa. Prog. Electromagnet. Res. B **15**, 113–131 (2009)

Chapter 8
Return Loss and Bandwidth Enhancement Using Back Fire Microstrip Patch Antenna

Puran Gour and Ravi Shankar Mishra

Abstract In this paper, we investigated and proposed a design of backfire microstrip patch antenna with improved bandwidth and return loss. The proposed antenna was designed and fabricated with given set of parameters with suitable design equations and method of movement approach. The simulated result shows that bandwidth is around 37.5 %, return losses is approximately −45 dB, VSWR < 1.5 and directivity is 8 dBi. The hardware fabrication and testing is also performed and validation results are close to simulated results. The proposed antenna is suitable for S and C band.

Keywords Backfire microstrip patch antenna • VSWR • Bandwidth • Directivity • Coaxial/Probe feeding • Flame retardant 4 (FR-4)

8.1 Introduction

Antenna is one of the important elements in the RF system for receiving and transmitting the radio wave signals from and into the air as the medium. If antenna is improperly designed, the signal from the RF system may not be transmitted and no signal can be detected by the receiver. Different types of antennas have been designed to cater with variable application and suitable for their needs. The backfire antenna was originally described in 1959 by H.W. Ehernspeck and much simplified version called the short backfire antenna (SBA) [1, 6]. A SBA is a type

P. Gour (✉)
Aisect University, Bhopal, India
e-mail: purangour@rediffmail.com

R. S. Mishra
Sagar Institute of Science and Technology, Bhopal, India
e-mail: ravishankarmishra@rediffmail.com

R. Maringanti et al. (eds.), *Proceedings of Ninth International Conference on Wireless Communication and Sensor Networks*, Lecture Notes in Electrical Engineering 299, DOI: 10.1007/978-81-322-1823-4_8,

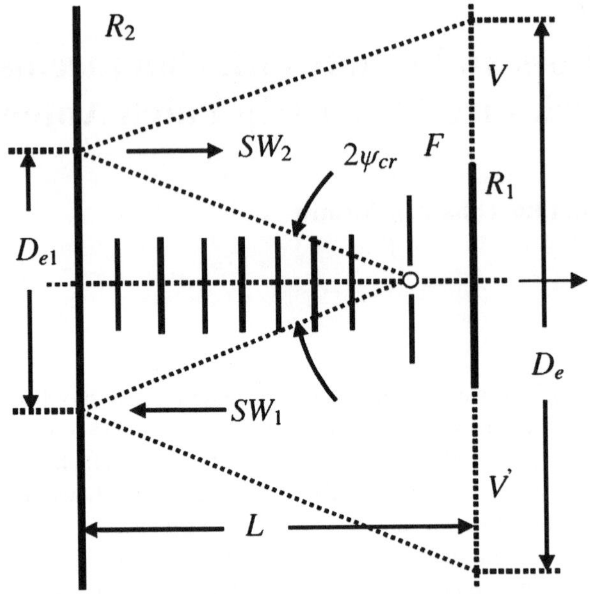

Fig. 8.1 Basic geometry of back fire antenna

of a directional antenna characterized by high gain, relatively small size, and narrow band [2–6]. The back fire microstrip patch SBA is superior over existing designs in its reduced size and mass while maintaining comparable performance [3, 4]. The backfire antenna can be designed by placing a big reflector at the open end perpendicular to its axis. The geometry of the backfire antenna is shown in Fig. 8.1. It consists of a source *F*, Surface wave structure *S*, and two parallel disk reflectors: *R*1 as small reflector and *R*2 as big reflector. Which reflects the surface wave SW2 toward the small reflector *R*1, where it is radiated from the antenna aperture *VV′* into the space due to which the radiation of the antenna is directed in inverse direction in comparison with the radiation of the ordinary end-fire antenna used as a backfire antenna prototype. So it is called Backfire Antenna [1, 4].

8.2 Feeding Techniques

There are four feeding techniques that can be used while designing the backfire microstrip patch antenna [7]. These are coaxial probe/probe coupling, microstrip feed, proximity (Electromagnetically)-coupled microstrip antenna, and aperture couple microstrip antenna feed. In this design, the coaxial probe and inset feed are used due to it simple fabrication and easy impedance matching (Fig. 8.2).

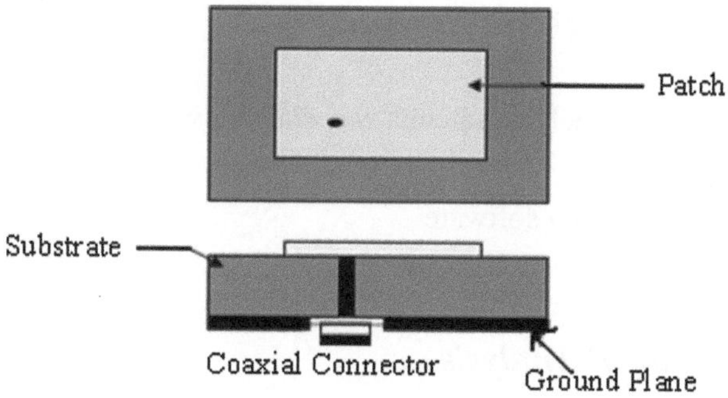

Fig. 8.2 Coaxial probe feed

Table 8.1 Comparison of different feeding methods

Characteristics	Line feed	Coaxial feed	Aperture coupled	Proximity coupled
Configuration	Coplanar	Non planar	Planar	Planar
Spurious feed radiation	More	More	More	More
Polarization purity	Good	Good	Excellent	Poor
Ease of fabrication	Easy	Soldering and drilling needed	Poor	Poor
Reliability	Better	Poor due to soldering	Good	Good
Impedance matching	Easy	Easy	Easy	Easy
Bandwidth (%)	2–5	2–5	2–5	13

8.2.1 Coaxial Feed

Coaxial probe is a coupling of power through a probe. A typical SBA is using *N* type coaxial connector. The coaxial probe is attached to the backside of the printed circuit board, and the coaxial center conductor after passing through substrate is soldered to the patch metallization. The location of feed point is determined for the given mode so that the best impedance match is achieved. It has narrow bandwidth and difficult to manufacture, especially for thick substrates [7].

8.2.2 Comparative Analysis of Different Feeding Methods

The different feeding techniques for microstrip patch antenna are available. The selection of feeding technique can be done by trade-off between different characteristics parameters of our design requirement. A comparative analysis is given in Table 8.1.

By comparing various feeding technique coaxial feed technique gives better impedance matching and easy to fabrication as compare to others, and it allows independent optimization of the feed and radiating parts of the antenna due to the metal ground plane placed between them.

8.3 Designing of Backfire Antenna

Designing of proposed backfire antenna was done in two stages:

1. Mathematical Analysis
2. Antenna design by using software

8.3.1 Mathematical Analysis

Mathematical analysis is necessary for knowing the exact dimension of the patch to be designed. By performing mathematical analysis width and length of the patch, ground plane and reflectors can be calculated. The bandwidth of the SBA is inversely proportional to the square root of substrate dielectric constant (ε_r). Substrate thickness is another important design parameter. Thickness of the substrate increases the fringing field at the patch periphery. It also gives lower quality factor and so higher bandwidth. The low value of dielectric constant increases the fringing field at the patch periphery, and thus increases the radiated power. A small value of loss tangent is always preferable in order to reduce dielectric loss and surface wave losses [7].

There are three essential parameters which is required for mathematical analysis of an antenna:

- Frequency of operation (*f*): The resonant frequency of the antenna must be selected appropriately.
- Dielectric constant of the substrate (ε_r): The dielectric material selected for proposed design for microstrip patch and ground plane is FR-4, which has a dielectric constant of 4.3. Also an effective dielectric constant (ε_{eff}) must be obtained in order to account for the fringing and the wave propagation in the line.
- Height of dielectric substrate (*h*): The height of the dielectric substrate is 1.5 mm for a proposed design.

Mathematical Analysis: [7]

$$w = \frac{c}{2f\sqrt{\frac{(\varepsilon_r+1)}{2}}} \tag{3.1}$$

$$\varepsilon_{eff} = \frac{\varepsilon_r+1}{2} + \frac{\varepsilon_r-1}{2}\left[1+12\frac{h}{W}\right]^{-\frac{1}{2}} \tag{3.2}$$

$$\Delta L = 0.412h\frac{(\varepsilon_{reff}+0.3)\left(\frac{W}{h}+0.264\right)}{(\varepsilon_{reff}-0.258)\left(\frac{W}{h}+0.8\right)} \tag{3.3}$$

$$L = \frac{c}{2f\sqrt{\varepsilon_{\mathrm{eff}}}} - 2\Delta L \quad (3.4)$$

$$L_0 = L + 6h \quad (3.5)$$

$$W_0 = W + 6h \quad (3.6)$$

where,

C Velocity of light
f Operating frequency
ε_r Permittivity of the dielectric
ε_{eff} Effective permittivity of the dielectric
W Patch's width
L Patch's length
h Thickness of the dielectric
L_0 Length of ground plate
W_0 Width of ground plate

Effects of Substrate

For designing of an antenna, substrate plays a very important role and the selection of substrate depends on various factors like: [7]

- The bandwidth of patch antenna is directly proportional to the substrate thickness (h) and inversely proportional to the square root of substrate dielectric constant (ε_r).
- Thick substrate increases the edge field at the patch periphery and thus increases the radiated power.
- A small value of loss tangent is always preferable in order to reduce dielectric loss and surface wave loss, and it increases the efficiency of antenna.

8.3.2 Antenna Design by Using IE3D Software

The proposed antenna designs consist of a pair of reflectors and patch excitation slot. The shape of main reflector is rectangular and another is circular. The excitation slot and main reflector are separated by FR4 substrate with a distance of 1.5 mm and the distance between main reflector and subreflector is 53.3 mm, that is, approximately equal to 0.5λ. The design parameter values which are given in Table 8.2, the dimension of proposed antenna are calculated by using equation (3.1)–(3.6). The return loss, directivity, and the radiation pattern can be obtained by using the powerful EM simulator tools IE3D (version 9.0).

Table 8.2 Values from mathematical calculation for proposed antenna

S. No	Parameters	Dimension of proposed antenna
1	Resonant frequency (f)	6 GHz
2	Dielectric constant (ε_r)	4.3
3	Height of substrate (h)	1.5 mm
4	Loss tangent of FR4	0.019
5	Width of rectangular patch (W)	15.357 mm
6	Length of rectangular patch (L)	11.5128 mm
7	Width of ground plane (W_0)	24.357 mm
8	Length of ground plane (L_0)	20.5128 mm
9	Diameter of subreflector (d)	48 mm
10	Height of subreflector (H_r)	53.3 mm
11	Feed location: X_f (along length) Y_f (along width)	$X_f = 4.75$ mm $Y_f = -5.35$ mm

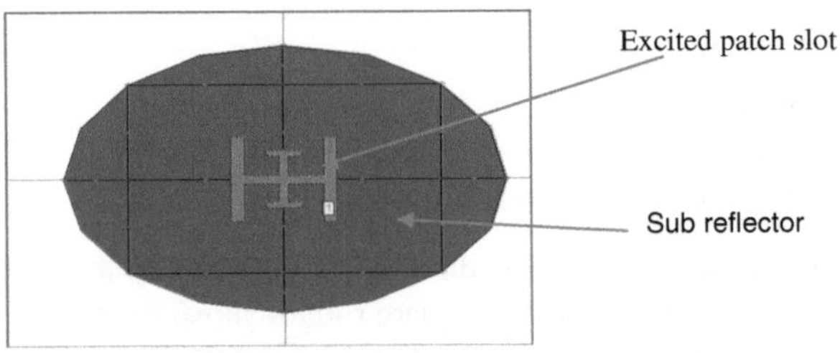

Fig. 8.3 Top view of suggested backfire antenna

8.4 Simulation and Results

The very powerful simulation tools was used to simulated the proposed design of antenna by modeling equation with IE3D software, which uses the Method of Moments (MOM) approach. The desired structure of the above-mentioned model is as shown in Fig. 8.3.

The geometry consists of two reflectors, one is rectangular and another is circular in between them and an excitation patch slot is connected.

The 3D view of the proposed antenna is as shown in Fig. 8.4.

Figure 8.5 shows that the return loss(S11) of the antenna is −45 dB at the center frequency of 3.73 GHz. The bandwidth obtained from the return loss result is 37.5 %, which is quite high as compared to the reported result [6].

Figure 8.6 dipicted that the VSWR is less than or approxitely equal to 1.5 at 3.23–4.73 GHz. This shows that the maximum power is transmitted from source end, thus there is less mismatch at source end.

Figure 8.7 shows the directivity of proposed antenna is 8dBi at 3.35 GHz, Which shows the proposed antenna is directional antenna.

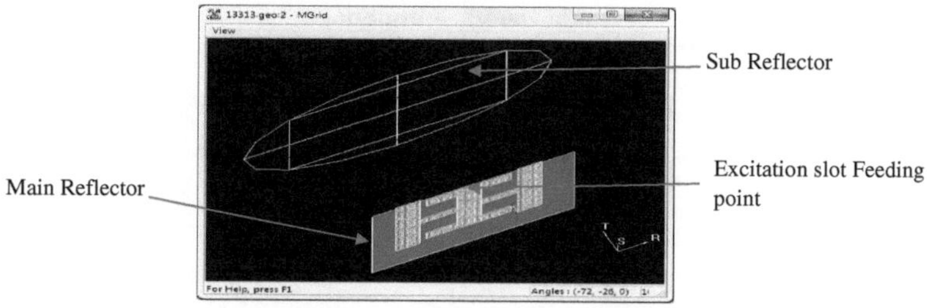

Fig. 8.4 3D view geometry of backfire antenna

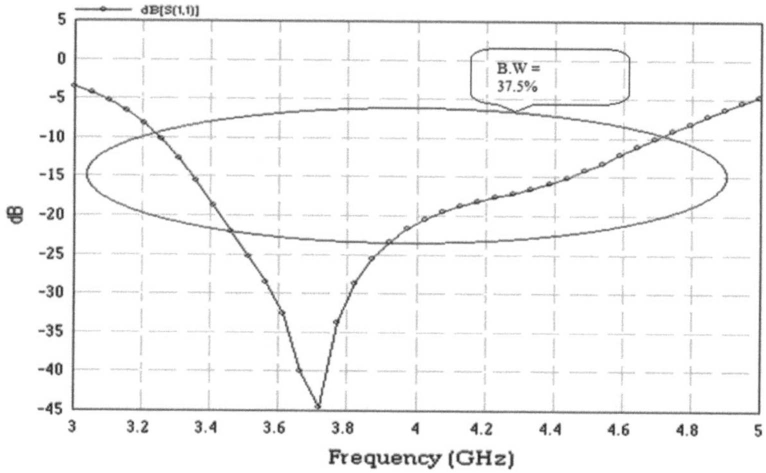

Fig. 8.5 S11 (Return loss) of proposed antenna

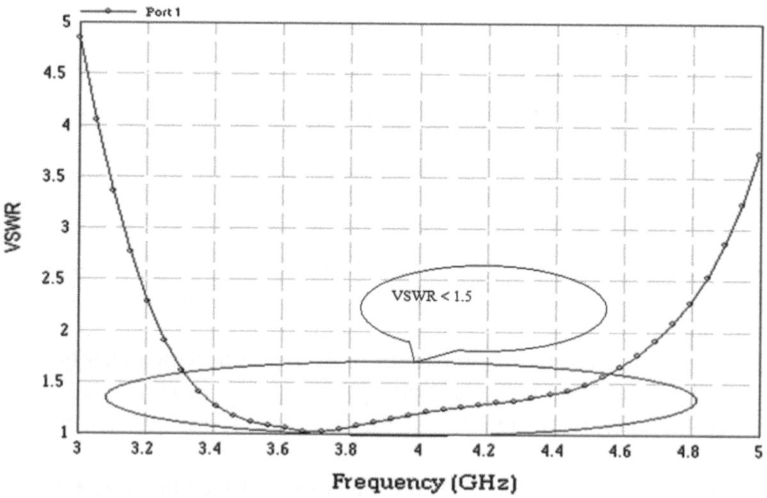

Fig. 8.6 VSWR curve for proposed antenna

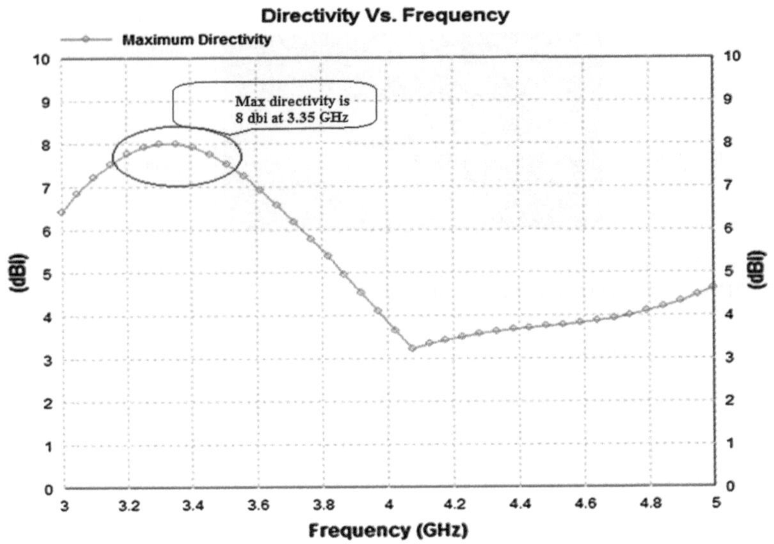

Fig. 8.7 Directivity versus frequency curve for proposed antenna

Fig. 8.8 Radiating efficiency of proposed antenna

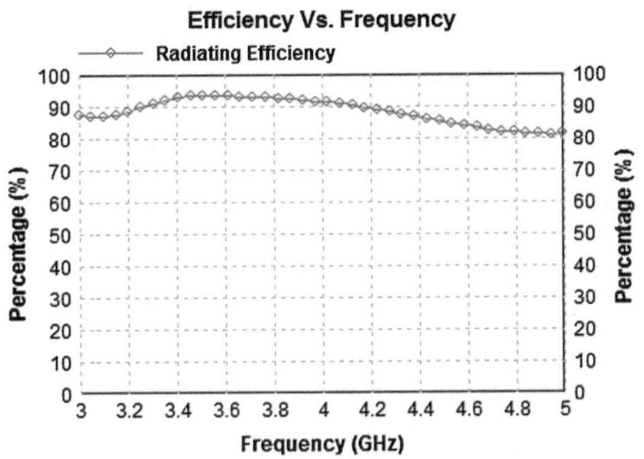

Figure 8.8 shows the proposed antenna radiation efficiency is approximately 90 % at 3.4–4 GHz, Which result that maximum radiation is obtained from proposed antenna.

Figure 8.9 depicts the radiation pattern in 3D domain, which clearly shows that the field radiation is only in forward direction and there is very small radiation at backside so the maximum radiation in forward direction is stable throughout the whole operating band.

The simulated result of proposed antenna is summarized in Table 8.3.

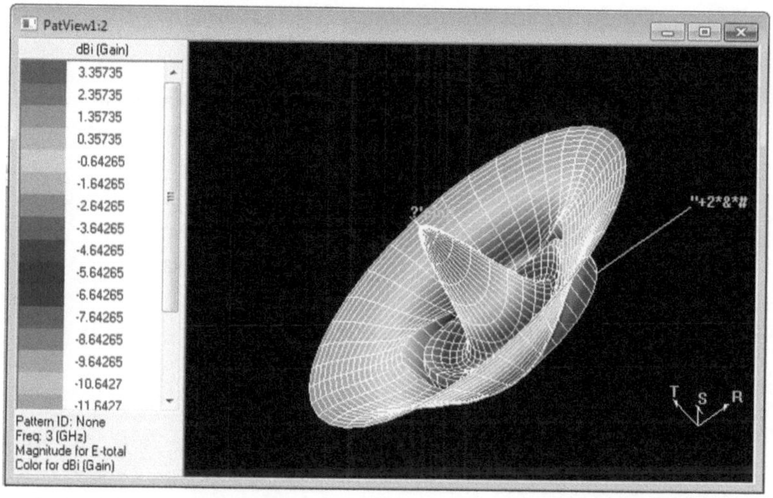

Fig. 8.9 3D Radiation pattern of proposed antenna

Table 8.3 Summary of simulated result

S.No.	Parameters	Simulated result
1	S11 Parameter	−45 dB at 3.72 GHz
2	Bandwidth	37.5 % between (3.23–4.73) GHz
3	VSWR	<1.5 between (3.32–4.52) GHz
4	Directivity	8 dBi at 3.35 GHz

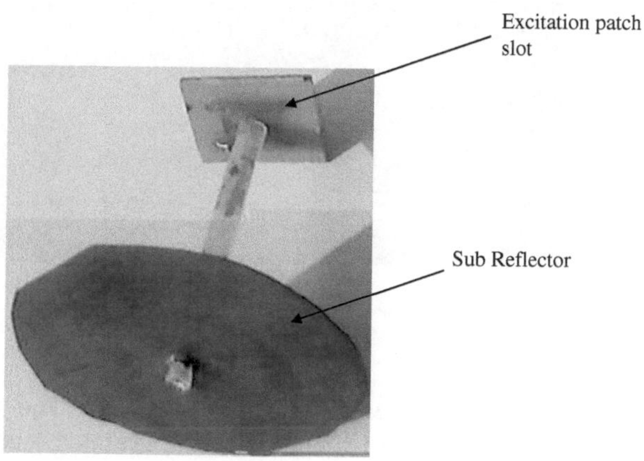

Fig. 8.10 Side view of proposed antenna

8.5 Hardware Implementation

Figures 8.10 and 8.11 show that side view and bottom view of proposed antenna, which is fabricated with glass epoxy material(FR4) ($\varepsilon_r = 4.3$) with both side copper plated. The thickness of FR4 is 1.5 mm and distance between main

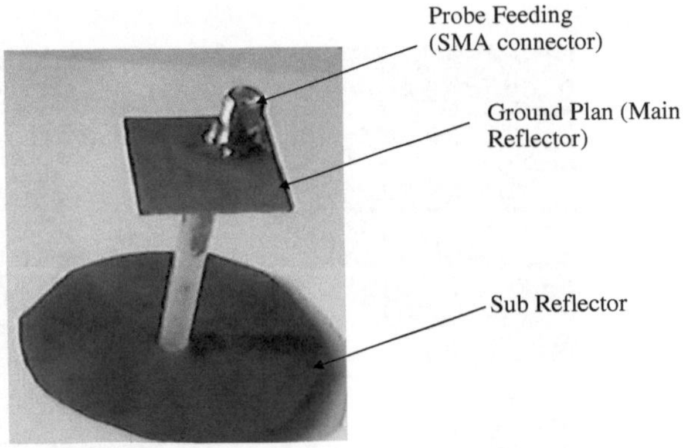

Fig. 8.11 Bottom plane of proposed antenna

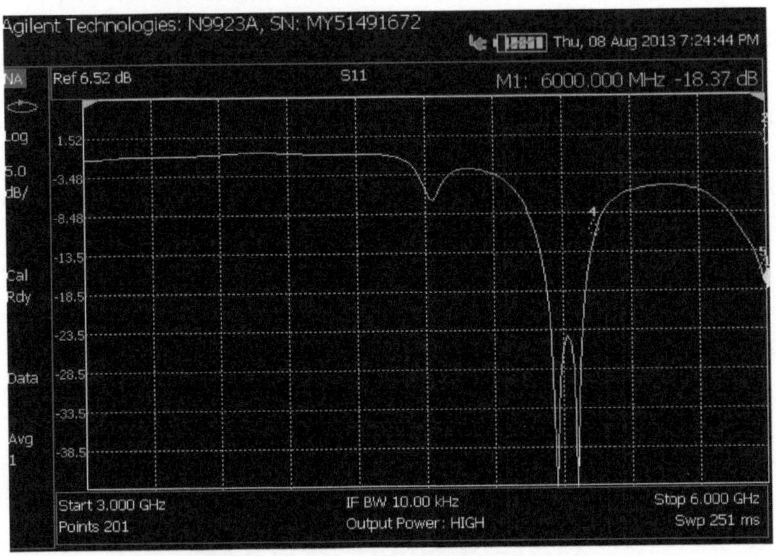

Fig. 8.12 Return loss for backfire antenna

reflector and subreflector is around 53.3 mm. The subreflector is made of copper plate. The patch excitation slot and ground plane(main reflector) are 1.5 mm apart (Fig. 8.12).

The hardware validation was done by an industrial measuring device that is network analyzer (Agilent N9923A). The testing result shows that the return loss(S11) is approximately −40 dB, which is closed to our simulated result. So this hardware can be used in S and C band .

8.6 Conclusions

The proposed design was successfully simulated on powerful simulator tools IE3D. The Simulated result shows that the return loss is approximately −45 dB, VSWR < 1.5, bandwidth is around 37.5 %, which is quite suitable for S and C band. The hardware was designed, fabricated, and tested by industrial measuring device network analyzer (Agilent N9923A). The hardware testing result shows that return loss is approximately −40 dB which is close to simulated result. So this hardware is suitable for S and C band and can be used for various wireless communications like WiMax, etc.

References

1. Kirov, G.S., Hristov, H.D.: Study of backfire antenna J. Microwave. Optoelectron. Electromagnet. Appl. **10**(1), 1–12 (2011)
2. Asad, M.J., Zafrullah, M., Islam, M.K., Amin, M.: Development of short backfire antenna fed by H-shaped excitation structure. In: 17th International Conference on Telecommunication in Pakistan, pp.449–454 (2010)
3. Qu, S.W., Li, J.L., Xue, Q., Chan, C.H., Li, S.:Wideband and unidirectional cavity—backed folded triangular bowtie antenna. IEEE Trans. Antenna propag. **57**(4), 1259–1263 (2009)
4. Kirov, G.S.: Design of short backfire antennas. IEEE Antennas propag. mag. **51**(6) (2009)
5. Li, R.L., Thompson, D., Tentzeris, M.M., Laskar, J., Papapolymerou, J.: A circular polarized short backfire antenna excited by an unbalance—fed cross aperture. IEEE Trans. Antenna propag. **54**(3), 852–859 (2006)
6. Li, R.L., Thompson, D., Tentzeris, M.M., Laskar, J, Papapolymerou, J.,: A new excitation technique for wide-band short backfire antennas. IEEE Trans. Antenna propag. 53(7), 2313–2320 (2005)
7. Balanis, C.A.: Antenna Theory Analysis and Design, Third Edition. Wiely, India, pp. 811–855 (2012)

Chapter 9
Dynamic Spectrum Sensing in Cognitive Radio Networks Using Compressive Sensing

Neeraj Kumar Reddy Dantu

Abstract In this paper, we propose a compressive sensing-based dynamic spectrum sensing algorithm for a cognitive radio network. The algorithm assumes the knowledge of initial energies in occupied channels and by using a number of wideband filters as a sensing matrix and $l-1$ minimization-based dynamic detection algorithm, iteratively determines the change in occupancy of channels. The advantages of such an algorithm include reduced number of filters than in previously used algorithms and a better performance at low SNRs. The performance of the algorithm is studied by varying different parameters involved and the results are shown. We demonstrate that the algorithm is effective and robust to noise.

Keywords Dynamic spectrum sensing • Cognitive radio • Compressive sensing application • Dynamic detection • Cognitive radio network

9.1 Introduction

Recent studies on patterns of spectrum usage and congestion in spectrum usage have warranted the use of smart radios that detect holes in the spectrum for a more advantageous and efficient usage of the spectrum. Specifically, the study of general spectrum usage through fixed spectrum assigning policy revealed that temporal and geographical usage of spectrum varies from 15 to 85 % [1]. The advent of wireless mobile devices resulted in higher bandwidth requirements. Considering the sparse nature of spectrum usage, spectrum sensing and temporal allocation of spectrum without interfering with the owner of the bandwidth (Primary radio) is an efficient

N. K. R. Dantu (✉)
Department of Electronics and Communication Engineering, The LNM Institute of Information Technology, Rupa-ki-Nangal, Jaipur 303012, India
e-mail: neerajkumarlnmiit@gmail.com

R. Maringanti et al. (eds.), *Proceedings of Ninth International Conference on Wireless Communication and Sensor Networks*, Lecture Notes in Electrical Engineering 299,
DOI: 10.1007/978-81-322-1823-4_9,

way to overcome this problem. These "smart" radios are called cognitive radios. In a Cognitive Radio network, the unlicensed Secondary Radios sense the spectrum and detect unused channels assigned to license Primary Radios to use these channels for information transfer without interfering with usage of the channel by the Primary Radio thereby increasing the efficiency of spectrum usage.

Spectrum sensing [2, 3] is one of the most important aspects of cognitive Radio. The desired characteristics of a spectrum sensing cognitive radio described in [3] are given below.

1. Highly efficient detection of spectrum holes and their classification in presence of noise.
2. High spectral resolution of detection.
3. Estimation of Direction of Arrival (DOA) of interferences.
4. Time-frequency analysis.

In this paper, we will be dealing with the first of these characteristics which are efficient detection of spectrum holes and determination of their occupancy by primary user. Many increasingly efficient methods have been proposed to estimate the occupancy of the channels. The nature of the algorithm used depends on the scenario. Factors like number of antennas, directivity of antennas, and structure of the network effect the nature of algorithm along with hardware and software complexities of the radio node. Energy detector-based sensing, waveform-based sensing, cyclostationarity-based sensing, and interference-based sensing are discussed in [2] and [3]. In this paper, we will discuss Compressive sensing-based spectrum sensing for a specific scenario.

Compressive sensing is a rapidly developing field with a wide range of applications. The basic idea behind compressive sensing is that if we have a signal that is sparse, it can be recovered with overwhelming probability taking considerably less sampling measurements than the dimension of the signal [4]. As sparsity is an essential property for compressive sensing and also the goal of compressive sensing [5] is to reduce the computational complexity and hardware requirements of the system, we can conclude that usage of this area might be beneficial in cognitive radio applications.

The paper is organised as follows: Previous work on this topic is discussed in Sect. 9.2. A short introduction and relevant theory to Compressive sensing is given in Sect. 9.3. Problem statement and system model are discussed in Sect. 9.4. The proposed algorithm and issues related are discussed in Sect. 9.5. Simulations and results are presented in Sect. 9.6. Section 9.7 concludes the paper.

9.2 Previous Work

It has been shown that the use of Compressive sensing is useful in cognitive radio applications [6] uses compressive sensing to reduce the dimension of measured signal and a wavelet-based edge detection scheme to look for spectrum holes and classification of channels. In [7], decentralised compressive detection is used to detect the occupancy of channels rather than reconstructing the whole spectrum. In both the schemes, sparsity of the signal decides the dimension of sensing matrix [8] uses a centralized model with dynamic compressive sensing which detects only the most recent change in spectral occupancy. By assuming an optimal sensing interval, this method further increases the sparsity of the measured spectrum thus reducing the dimension of the sensing matrix which translates to the number of wideband filters. Major advantage of this method is that the sparsity is now a function of the changes in spectrum usage rather than the spectrum usage itself. However, the paper uses l2-minimization or least squares which is not an optimal algorithm for compressive sensing [9]. Furthermore, centralized sensing mechanisms involve fusion centre or central node collecting a lot of information periodically from each node to update the spatial spectral density which is a bottleneck [10]. l1-minimization has been established as the optimal algorithm for recovery of the signal in a compressive sensing model [11]. In our paper, we propose an algorithm that is loosely based on the dynamic spectrum sensing algorithm in [8], that uses the architecture of [7] and l1-minimization improving the efficiency of spectrum sensing and further reducing the number of wideband filters required for spectrum sensing. A brief introduction to theory of compressive sensing is provided in the next section.

9.3 Compressive Sensing

9.3.1 Compression

Normally, any basis matrix will be an orthogonal unitary square matrix of the dimension of the signal. The analysis and synthesis matrices are transposes of each other. In compressive sensing, the analysis and synthesis matrices are not orthogonal. The "sensing matrix," Ψ is rectangular and of lesser dimension than the square orthogonal matrix. Specifically, given a signal $x \in R^N$, we consider measurement systems that acquire M linear measurements. We can represent this process mathematically as

$$y = \Psi x \tag{9.1}$$

where Ψ is an $M \times N$ matrix and $y \in R^M$. The matrix Ψ represents a dimensionality reduction, i.e., it maps a signal in R^N, where N is generally large, into R^M, where $M \ll N$.

There are conditions that both the signal and sensing matrix should satisfy in order to ensure a complete recovery of the signal with high probability. x should be compressible, i.e., it's sorted coefficient magnitudes should exhibit power law decay and sparse, i.e., $\|x\|_0 \leq K$ where $K \ll N$ where N is the actual dimension of the signal. Taking this sparsity of the signal into account, the theory of compressive sensing imposes conditions on sensing matrix Ψ too. Two most important properties that sensing matrices should satisfy are the null space property (NSP) and restricted isometry property (RIP) [12]. The next question that arises is what can be the dimensions of this sensing matrix or how far can we compress the signal. The dimension of the sensing matrix depends on the sparsity of the signal as well as the probability of recovery required. This means that if the number of nonzero coefficients of the signal increase, the probability of exact recovery decreases for same dimensions of sensing matrix. Probability of recovery also decreases with decrease in dimensions of sensing matrix. Specifically, we should choose $M > CK \log(N/K)$ [4, 12], where C is a constant, K is the sparsity, and N is the actual dimension of the signal.

Now that we have established what properties the signal and the sensing matrix, we need to verify them in our case. The spectrum occupancy is sparse and if we arrange the channels having the highest energy to the lowest, the coefficients will obey a power law decay. So, we deem the signal in frequency domain to be compressible and sparse. Now, we need a Sensing matrix that obeys the above-mentioned properties and is of a dimension that is convenient to our application. Designing a sensing matrix is very difficult. Showing that these matrices have these properties is also a gargantuan task. It was discovered that RIP and incoherence can be achieved by simply choosing a random matrix as sensing matrix. Many papers deal with Gaussian, more specifically, sub-Gaussian random matrices, or Bernoulli random matrices [9, 12] and discuss their usage in different compressive sensing applications. In the words of Richard Baraniuk, a pioneer in this field, if you generate a random matrix in MATLAB using its function, there is a high probability that it will have the properties we need.

9.3.2 Recovery

There are many ways to recover the signal depending on the error of approximation and the extent of compression needed. To control the degree of approximation, we employ cost functions. The computational complexity of signal recovery is governed by these cost functions [9]. The most logical cost function taking sparsity into account seems to be the cardinality or the zero norms. The optimisation problem can be mathematically formulated as follows.

$$\begin{aligned} &\underset{x \in C^N}{\text{minimize}} \ ||x||_0 \\ &\text{subject to } \Psi x = y \end{aligned} \tag{9.2}$$

If we do that, the problem of signal recovery will become NP hard [5]. $l-1$ and $l-2$ norms seem to be the alternatives. They can be handled as Linear programming problems which are computationally much less complex than $l-0$ minimization problem. However, $l-1$ norm minimization has been shown to be much more efficient than $l-2$ minimization.

We can understand why $l-2$ norm fails in the scenario if we look at the recovery problem of compressive sensing geometrically. If we consider a sparse signal $x \in R^N$, We can imagine an N dimensional coordinate axes in which the signal is nonzero in only some directions and its alignment is random. In signal recovery, the translated null space of the sensing matrix where the signal solution will lie is also randomly aligned due to the randomness in the sensing matrix. The process of $l - 2$ minimization geometrically means that a unit sphere is placed at the origin and is blown up until it reaches the null space of the sensing matrix. This solution might not be sparse or close to the solution which is clearly randomly aligned along few directions [9]. In case of $l - 1$ norm minimization, the unit ball is pointy and not spherical. Thus, there is a much greater probability of recovery evolving the $l - 1$ norm ball. The optimization problem is given below.

$$\begin{aligned} &\underset{x \in C^N}{\text{minimize}} \ ||x||_1 \\ &\text{subject to } \Psi x = y \end{aligned} \tag{9.3}$$

The most general methods involve greedy algorithms [13] and convex optimization algorithms [14]. While greedy algorithms include matching pursuit, orthogonal matching pursuit, and their variations, convex optimization algorithms use linear or cone programming.

9.4 System Model and Problem Statement

Let us consider a cognitive radio network that has a total bandwidth of W Hz divided among N cognitive radios. As stated in [7], it may be an ad hoc network where the cognitive radios share the whole bandwidth depending on the need or it may be a secondary network trying to use licensed bandwidth of primary radios. Furthermore, a centralized Fusion centre or a decentralized architecture can be used. In a centralized mechanism, the fusion centre or the central node collects spectral occupancy information from each node and dissipates the information whenever needed. As discussed earlier, this causes a bottleneck at the fusion centre. Decentralization can be achieved through partial or complete sharing of spectral information with each other between cognitive radios [2, 7] suggest use of a dedicated frequency control channel using cognitive radios which coordinate among themselves. The cognitive radios exchange spectral information and arrive at a consensus on which channel to use using this control channel. This will solve issue of hidden terminal problem which is a major obstacle for a decentralised network. Effects of fading or faulty detection can be reduced by using fault tolerance algorithms [15]. However, it is evident that exclusion of a fusion centre results in an increased hardware complexity of each cognitive radio node. This is a question of trade-off between repeated transmissions to fusion centre and possible congestion to hardware complexity and therefore cost of each cognitive radio node. The algorithm that we propose will work for centralized as well as decentralised networks.

We use the system described in [7]. We assume that each node has a wideband antenna that gives a wideband signal $x(t)$. Each node has K wideband filters where $K \ll N$ depends on the sparsity of the channel. These K filters are such that $\Psi_{ki} = H_k(f_i)$, $k = 1, 2, 3, \ldots, K$ $i = 1, 2, 3, \ldots, N$. Here, $H_k(f)$ represents transfer function of kth filter and f_i represents the ith channel. The time-domain wideband signal is passed through the filters and the energy of the output signal of each filter is then measured to get the $K \times 1$ energy vector Y.

$$Y_k = x_k x_k^H \quad k = 1, 2, 3, \ldots, K \tag{9.4}$$

$$Y = [Y_1 \quad Y_2 \quad Y_3 \quad \ldots Y_K] \tag{9.5}$$

Now, if we denote E_i as the energy of the signal in the ith channel, then

$$E_i = \int_{f_i - B/2}^{f_i + B/2} \mathcal{F}(x(t))\mathrm{d}f \tag{9.6}$$

If we know the filter transfer function $H_k(f)$, the output of kth filter can be represented as

$$Y_k = \sum_{i=1}^{N} (|H_k(f_i)|)^2 E_i \quad k = 1, 2, 3, \ldots, K \tag{7}$$

We can represent this by the following equation:

$$Y = \Psi E \tag{9.8}$$

where

$$\Psi = \begin{bmatrix} (|H_1(f_1)|)^2 & (|H_1(f_2)|)^2 & \cdots & (|H_1(f_N)|)^2 \\ (|H_2(f_1)|)^2 & (|H_2(f_2)|)^2 & \cdots & (|H_2(f_N)|)^2 \\ \vdots & \vdots & \vdots & \vdots \\ (|H_K(f_1)|)^2 & (|H_K(f_2)|)^2 & \cdots & (|H_K(f_N)|)^2 \end{bmatrix} \tag{9.9}$$

and Y is the K dimensional compressed energy vector.

In a dynamic environment, where E and consequently Y change if a cognitive radio starts or stops using a channel, we assume that the cognitive radio has an initial Y and E. After a fixed interval based on the frequency of changes in spectrum usage, the cognitive radio compressively measures the spectrum again giving Y_{pres} for the present iteration. Let E_{prev} and Y_{prev} be the energy vector and compressed energy vector in the previous measurement. The sensing matrix Ψ, Y_{pres}, E_{prev}, and E_{prev} are used by the cognitive radio or the fusion centre to extract E_{pres}. The algorithm loosely based on [8] is described in the next section.

9.5 Proposed Algorithm for Dynamic Detection

As described in the end of previous section, the goal is to estimate E_{pres} using E_{prev}, Y_{prev}, Y_{pres}, and Ψ. We obtain ΔY by subtracting Y_{prev} from Y_{pres}.

$$\Delta Y = Y_{\text{pres}} - Y_{\text{prev}} = \Psi \left(E_{\text{pres}} - E_{\text{prev}} \right) \tag{9.10}$$

So, the change in already sparse energy vector E will be reflected in change in Y. In order to recover E_{pres}, we cast the following optimisation problem.

$$\begin{aligned} &\underset{\Delta E}{\text{minimize}} \ ||\Delta E||_1 \\ &\text{subject to } \Psi \Delta E = \Delta Y \end{aligned} \tag{9.11}$$

where $\Delta E = E_{\text{pres}} - E_{\text{prev}}$. We already know E_{prev}. So, E_{pres} can be calculated as

$$E_{\text{pres}} = E_{\text{prev}} + \Delta E \tag{9.12}$$

Here, ΔE represents the change in occupancy of the spectrum. By doing this, we have reduced the sparsity requirement of the energy vector to be recovered to the change in number of channels used. A simple Energy detection scheme on the recovered energy vector can be used to determine the change in channel usage.

After a time period of sampling after which a change in spectrum usage is expected, the cognitive radio must repeat the above algorithm after updating Y_{prev} and E_{prev}. The updated values of Y_{prev} and E_{prev} are calculated as shown below.

$$Y_{\text{prev}}(n+1) = Y_{\text{pres}}(n) \tag{9.13}$$

$$E_{\text{prev}}(n+1) = E_{\text{pres}}(n) \tag{9.14}$$

where n is the iteration number. Then, the cognitive radio uses the spectrum sensing mechanism described earlier to estimate the present compressed energy vector Y_{pres}. The optimisation problem cast in (9.11) can be solved using standard optimisation tools such as cvx or l1-magic. The step-by-step process of the algorithm is shown below.

Algorithm 1 Algorithm for Dynamic Detection

Input:

E_{prev}, Y_{prev} and Ψ.

Output:

E_{pres}.

1: Obtain the compressed energy vector Y_{pres} from wideband filter based compressive sensing mechanism.
2: Calculate $\Delta Y = Y_{pres} - Y_{prev}$.
3: Estimate optimal ΔE by solving the optimisation problem in 11.
4: Compare ΔE with threshold to determine the change in occupancy of channels.
5: Calculate E_{pres} using 12.
6: Update E_{prev} and Y_{prev} using 13 and 14.
7: Go to step 1 after a sampling time period of spectrum estimation.

9.5.1 Limitations and Advantages

The obvious limitation of the algorithm is its increased computational complexity. If a fusion centre-based network architecture is chosen, only the fusion centre bears the additional burden of this increased complexity. The individual cognitive radio's hardware complexity requirement decreases due to decrease in number of wideband filters required when compared to the algorithm used in [7]. In addition to the decreased number of filters, the length of wideband filters is much less than the traditional narrowband filters. The dynamic nature of the algorithm also brings certain shortcomings to the algorithm. The error of detection of an unoccupied channel or nondetection of occupied channel will propagate through the next iterations as described in [8]. As discussed earlier, Fault tolerating cooperative sensing mechanisms can be used to detect and rectify errors of individual cognitive radio.

An important advantage of the algorithm is its ability to detect change in occupancy of multiple channels compared to the dynamic algorithm in [8] which assumes a change in occupancy of one channel only. This reduces the sampling rate or time interval of each iteration. But, this also increases the required number of filters as we will see in the next section. The sampling rate of spectrum sensing is a drawback of the algorithm compared to static spectrum sensing algorithms. The cognitive radio needs to continuously sense the channels after a sampling period even if it has no intention of transmitting information. This is one aspect of the algorithm in which a room for improvement can be seen. Another important issue is the scalability of the network. Networks often include large number of nodes. In a network with no capability of dynamic spectrum sensing, the increased number of nodes will result in proportional increase in the number of wideband filters. Change in channel occupancy is the only factor that controls the number of wideband filters in our algorithm. While it is true that the change in channel occupancy also increases with increasing number of nodes in the network, this increase is very less compared to the increase in number of nodes. Thus, the additional number of wideband filters required for increased number of nodes is much less than that of a network with static spectrum sensing mechanism. Moreover, we can control the hardware requirement by changing the sampling interval. For example, if a certain amount of nodes are added to the network, the addition of wideband filters can be avoided by decreasing the sampling time period so that the average change in channel occupancy per time period remains same as before the addition. But, this will increase the power usage of the network. So, it is important to strike a balance between number of wideband filters and sampling period to avoid inefficient functioning. Finally, it has been shown that Compressive Sensing-based mechanisms are highly tolerant to noise [4, 5]. So, the algorithm is highly tolerant to noise which is a highly desirable property of any spectrum sensing mechanism. As we can see, the advantages of the algorithm outweigh the limitations of its usage. The advantages and behavior of the algorithm with respect to change in different parameters are studied in the next section.

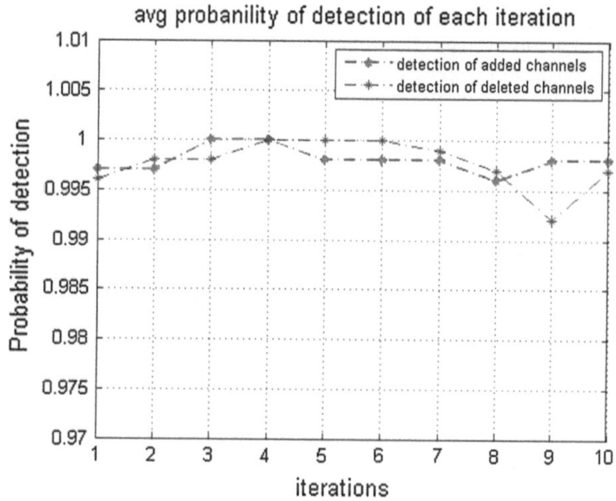

Fig. 9.1 Probability of detection versus iteration number for SNR = 2 dB and ratio of compression = 30 %

9.6 Simulations

In this section, we will discuss the performance of the algorithm by interpreting the results obtained from several simulations. We evaluate the performance of our dynamic detection scheme with respect to parameters including SNR, ratio of compression and change in channel occupancy. In the simulations, we consider a cognitive radio network of 50 cognitive radios sharing a wideband spectrum containing 50 channels of equal bandwidth. The noise is assumed to be Additive White Gaussian. Each cognitive radio has K wideband filters as described and they either execute the whole algorithm themselves or transmit the sensing information to the fusion centre which will determine the spectral occupancy at each iteration.

The results of a sample simulation are shown in Fig. 9.1. The figure shows the probability of detection of change in spectrum usage of 1 channel with a maximum occupancy of 11 channels in the total 50 channels. The SNR is 2 dB and the number of filters is 15(ratio of compression = 30 %). As we can see, there is no change of probability of detection whether the total occupancy of spectrum is 1 or 10. This proves that the efficiency of the algorithm depends on the change in occupancy of channels reflected in ΔE and not on the actual occupancy reflected in E in each iteration.

Figures 9.2 and 9.3 show how the average probability of error of detection of change in spectral occupancy varies with change in number of occupied channels and ratio of compression, respectively. In the simulation for Fig. 9.2, the ratio of compression (30 %) and SNR (3 dB) are kept constant and the number of channels added or deleted is varied from 1 to 6. As it can be seen, the probability of error increases with increase in number of channels added or deleted as it represents the sparsity of ΔE. In order to detect more changes, the number of filters, i.e., the ratio of compression needs to be increased. This behavior can be seen in Fig. 9.3. In the

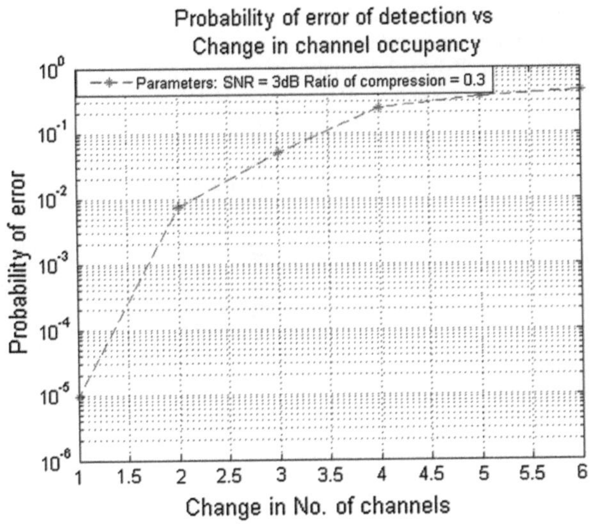

Fig. 9.2 Probability of error of detection versus change in occupancy of channels

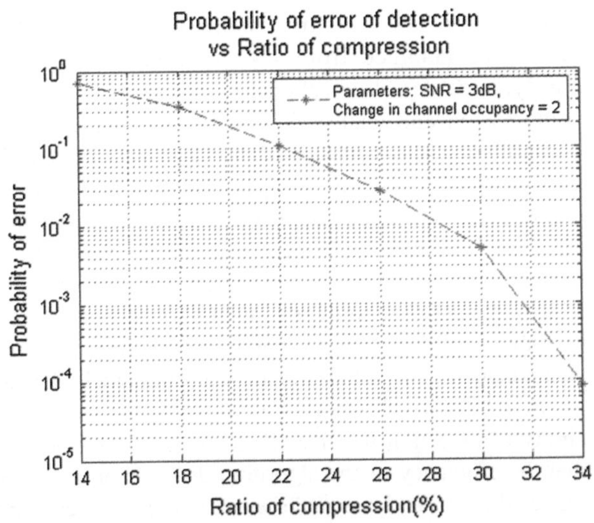

Fig. 9.3 Probability of error of detection versus ratio of compression

simulation, the change in channel occupancy (2) and SNR (3 dB) are kept constant and number of filters are varied. We see a significant change in probability of error by increasing the ratio of compression from 16 to 34 %. Figure 9.4 shows how the average probability of error of detection of change in spectral occupancy changes with respect to varying SNR. The ratio of compression is fixed to 30 % and change in spectral occupancy is fixed to be 1. The results suggest that the algorithm performs better than previously used algorithms in [7] and [8] even at low SNRs.

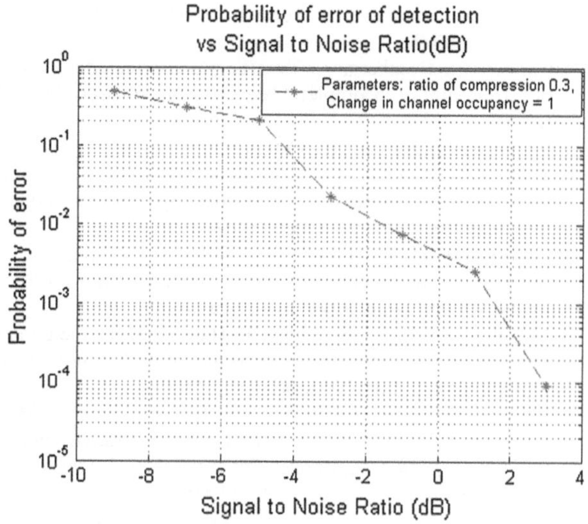

Fig. 9.4 Probability of error of detection versus SNR

9.7 Conclusion

In this paper, we proposed a compressive sensing-based algorithm for dynamic detection of occupancy of channels for a cognitive radio network. By using wide-band filters as a sensing matrix, a compressed energy vector was obtained by measuring their outputs and using previously measured or available original and compressed energy vectors, the energy vector that represented the energy in each channel in the present iteration was recovered using $l - 1$ minimization-based dynamic algorithm. The working and efficiency of the algorithm were studied by varying different parameters like ratio of compression (number of filters), change in number of occupied channels, and signal to noise ratio. The advantages of the algorithm like increased efficiency and decreased hardware complexity compared to other algorithms were discussed. Numerical results suggest that by choosing appropriate number of filters and time period of sampling, a very high probability of detection of change of spectrum occupancy can be obtained.

Acknowledgments I thank Dr. Pratik Shah for the encouragement and support he has given me. I thank my family who also supported me in the course of writing this paper.

References

1. Akyildiz, I.F., Lee, W.Y., Vuran, M.C., Mohanty, S.: Next generation/dynamic spectrum access/cognitive radio wireless networks: A survey. Comput. Netw. J. (Elsevier) **50**, 2127–2159 (2006)
2. Yucek, T., Arslan, H.: A survey of spectrum sensing algorithms for cognitive radio applications. IEEE Commun. Surv. Tutorials **11**(1), 116–130 (2009)

3. Haykin, S., Thomson, D., Reed, J.: Spectrum sensing for cognitive radio. Proc. IEEE **97**(5), 849–877 (2009)
4. Candes, E., Romberg, J., Tao, T.: Robust uncertainty principles: exact signal reconstruction from highly incomplete frequency information. IEEE Trans. Inf. Theor. **52**(2), 489–509 (2006)
5. Fornasier, M., Rauhut, H.: Compressive sensing (2010)
6. Tian, Z., Giannakis, G.: Compressed sensing for wideband cognitive radios. In: IEEE International Conference on Acoustics, Speech and Signal Processing, 2007, ICASSP 2007, vol. 4, pp. IV–1357–IV–1360. (2007)
7. Havary-Nassab, V., Hassan, S., Valaee, S.: Compressive detection for wide-band spectrum sensing. In: IEEE International Conference on Acoustics Speech and Signal Processing (ICASSP), 2010, pp. 3094–3097. (2010)
8. Yin, W., Wen, Z., Li, S., Meng, J., Han, Z.: Dynamic compressive spectrum sensing for cognitive radio networks. In: Proceedings of 45th Annual Conference on Information Sciences and Systems (CISS), 2011, pp. 1–6. (2011)
9. Baraniuk, R.: Compressive sensing. IEEE Sig. Process. Mag. **24**, 118–120 (2007)
10. Barbarossa, S., Scutari, G., Battisti, T.: Cooperative sensing for cognitive radio using decentralized projection algorithms. In: IEEE 10th Workshop on Signal Processing Advances in Wireless Communications, 2009. SPAWC '09, pp. 116–120. (2009)
11. Tropp, J.A.: Topics in sparse approximation. PhD thesis, University of Texas (2004)
12. Rauhut, H.: Compressive sensing and structured random matrices. Theor. Found. Numer. Meth. Sparse Recovery **9**, 1–92 (2010)
13. Tropp, J., Gilbert, A.: Signal recovery from random measurements via orthogonal matching pursuit. IEEE Trans. Inf. Theor. **53**(12), 4655–4666 (2007)
14. Boyd, S., Vandenberghe, L.: Convex Optimization. Cambridge University Press, New York (2004)
15. Krishnamachari, B., Iyengar, S.S.: Efficient and fault-tolerant feature extraction in wireless sensor networks. In: Proceedings of the 2nd International Workshop on Information Processing in Sensor Networks (IPSN 03). (2003)

Chapter 10
Scheduling Transmissions of Coexisting Wireless Body Area Networks Using Minimum Weight Match

Anagha Jamthe and Dharma P. Agrawal

Abstract Wireless Body Area Network (WBAN) is extensively being used for ubiquitous health care monitoring. Coexisting multiple WBANs could possibly suffer from high interference and thus an appropriate scheduling scheme is desirable to improve network throughput for interfering nodes. The received Signal to Interference and Noise Ratio (SINR) from the sensors in the overlapped region of two or more WBANs which use the same transmitting frequency drops. This decrease is an indicator of inter-WBAN interference. Hence, we propose a scheduling algorithm which assigns the node with minimum SINR to first available transmission slot and so on. Thus, the nodes which suffer maximum interference are given higher priority to transmit in the revised Time division multiple access (TDMA) schedules. The local coordinator of each WBAN exchange messages with the interfering WBAN's coordinator and then tries to suppress interference by creating a new TDMA schedule for interfering nodes. Results obtained using simulations are encouraging as they indicate that minimum weight match scheduling can improve the packet drop rate and thus enhance the network performance.

Keywords Base station (BS) • Bipartite graph • IEEE 802.15.6 • Local coordinator (LC) • Time division multiple access (TDMA) • Wireless body area network (WBAN)

A. Jamthe (✉) · D. P. Agrawal
Department of Electrical Engineering and Computing Systems 819D Old Chemistry, University of Cincinnati, Cincinnati, OH 45221-0300, USA
e-mail: jamtheas@mail.uc.edu

D. P. Agrawal
e-mail: agrawadp@mail.uc.edu

R. Maringanti et al. (eds.), *Proceedings of Ninth International Conference on Wireless Communication and Sensor Networks*, Lecture Notes in Electrical Engineering 299, DOI: 10.1007/978-81-322-1823-4_10, © Springer India 2014

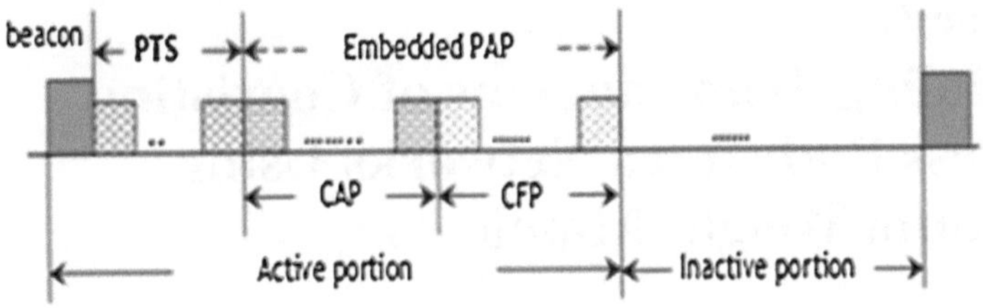

Fig. 10.1 Super-frame structure of IEEE 802.15.6 standard [2]

10.1 Introduction

The popularity of Wireless body area network (WBAN) in healthcare monitoring is increasing tremendously due to its potential applications such as monitoring elderly at home or athletes during their training and rehabilitation process [1]. WBAN is rapidly replacing wired healthcare devices due to its versatility such as ease of use and low power requirements. It is being explored extensively at its different layers such as Physical and Medium Access Control Layer (MAC layer) for different reasons such as efficient data transmission and energy consumption. IEEE 802.15 Task Group 6 has recently defined standards for the Physical and MAC layer in WBAN [2]. MAC layer is responsible for scheduling data transmissions and coordinate nodes' channel access in order to avoid possible collisions during data transmissions. An efficient scheduling scheme can achieve optimal tradeoff between acceptable amount of packet delay and maximum achievable network throughput.

Figure 10.1 shows the super-frame structure of the IEEE 802.15.6 standard [2]. This recently approved standard is specifically designed to support a WBAN. It is based on classifying the network traffic into 8 different categories according to user priority, which allows asymmetric flow of data [2]. The frame is divided into four periods namely: priority access period (PAP), contention access period (CAP), the contention free period (CFP), where the data transmission takes place and the idle period, where devices can enter into sleep mode. These periods are bounded within beacon intervals which are controlled by the coordinator. PAP supports the transmission of very high priority data such as emergency alarms in case of medical applications. Slotted ALOHA access method is used in PAP and CAP for robust data transmission and a Time division multiple access (TDMA) based scheme is used in CFP [2], which assigns unique transmission slots to the end devices in order to transmit periodical and QoS guaranteed data. TDMA based protocols are more useful to be employed in WBAN as they avoid energy wastage caused due to collisions and the nodes do not have to keep their radios continuously on for polling or assignment of next slot [3].

In a single WBAN, sensor devices transmit their sensed physiological data to the local coordinator (LC), which can be a PDA or a smart phone carried by person. Then, LC transmits received data to the medical server located at hospital or monitoring station, also known as base station (BS). It also schedules

Fig. 10.2 Inter-WBANs interference [4]

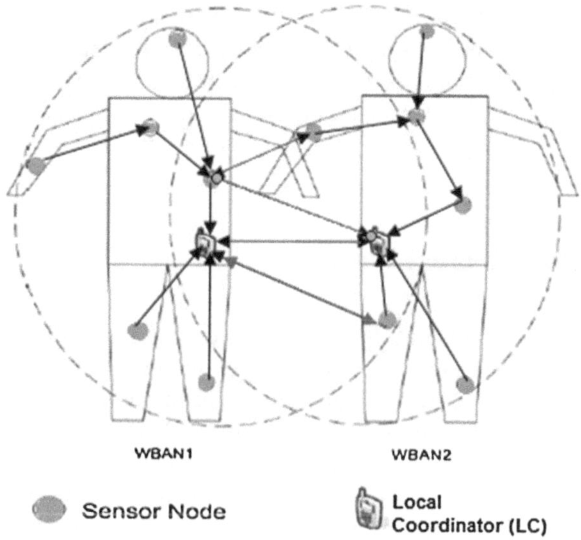

transmissions of end devices in CFP using TDMA and ensures that collision free transmission occurs inside the WBAN. In the places where multiple WBANs have to coexist in close proximity, such as health monitoring of various athletes in a sports arena and obtaining critical physiological parameters for numerous patients in a hospital, many sensor nodes can transmit concurrently using the same frequency (co-channel signals), which can cause a serious co-channel interference. Such a threat to the network performance could lead to loss of critical data, which can be even life threatening. Due to the distributed nature of WBANs, there is no central coordinator or BS to control transmissions from all coexisting WBANs. Thus, it is extremely critical to study the effects of inter-WBAN interference and propose ways to mitigate it.

When the two or more WBANs are within each other's transmission range, some of the sensors can cause serious co-channel interference as shown in Fig. 10.2. IEEE 802.15.6 standard suggest that graceful coexistence of up to 10 WBANs should be supported [2]. The co-channel signals arriving from undesired transmitters can cause network performance degradation. Each double ended arrow in Fig. 10.2 indicates possible interference between two adjacent WBANs which need to be scheduled carefully in order to avoid any loss of transmitted data.

Thus, in this paper, in order to harmonize the system when several WBANs co-exist in the vicinity, we propose a scheduling algorithm which uses minimum weight matching to mitigate interference. The interfering transmissions are scheduled based on minimum signal to interference and noise ratio (SINR). When two or more WBANs transmitting on the same frequency overlap, the received SINR by LC from these sensors falls, signals with SINR below 0 db are unacceptable, thus indicating interference [4]. The SINR of interfering neighbors can be lowered by coordinating broadcasted TDMA schedule from these sensors. An interleaved scheduled among multiple WBANs is prepared based on the matching minimum SINR value sensor id with first available transmission slot. Thus, the amount of packet loss is reduced and better network performance can be achieved. Although an overhead of exchanging SINR values and schedules exist, the system will perform better when WBANs coexist for considerable amount of time.

10.2 Related Work

Although extensive studies addressing scheduling in wireless sensor networks have been performed, challenges involved in scheduling rapidly-changing number of close by WBANs has not been addressed completely. Prabh et al. discuss opportunistic packet scheduling for WBAN [5] where a selected user gets an opportunity to transmit data in a particular slot so as to maximize the throughput of the entire network. But, this scheme has an underlying disadvantage that, the nodes have to keep listening to the medium, so that they can transmit data immediately as soon as a slot is allocated to them. This results in high energy consumption and is not suitable for ubiquitous functioning of WBANs.

In order to reduce the losses in data transmission, variable TDMA scheduling algorithm has been proposed by Tseilshchev et al. [3], which decides the order of transmissions in real time rather than keeping it fixed. Markov model is used as a means of formulating the optimization problem of maximizing the network throughput. However, the paper does not address the role of human movements to transmission losses.

Cheng et al. propose the use of Graph coloring algorithm for inter-WBAN scheduling [6]. Two approaches of random coloring and incomplete coloring in order to achieve optimal tradeoff between the convergence and high utilization of the channels have been proposed.

None of the above models use IEEE 802.15.6 standard which is designed especially for WBAN. In this paper we propose a MAC based scheduling protocol which reduces the interference in coexisting WBANs and is based on IEEE 802.15.6. The rest of the paper is organized as follows. Section 10.3 discusses our proposed system model as a weighted bipartite graph and formulizes the problem of scheduling scheme as matching of minimum SINR values received from interfering sensors to first available transmission slot. Section 10.4 discusses the computer simulations and results. Finally, the paper is concluded in Sect. 10.5.

10.3 System Model

WBAN can be modeled as a system of n sensor nodes capable of monitoring physiological parameters such as heart rate, body temperature, limb movement, blood pressure etc. This sensed data can be transmitted either in single hop or multi-hop to the

LC, which is a powerful device capable of collecting all the sensed information and transmitting it to the desired BS. LC is responsible for scheduling the intra-WBAN transmissions and resource allocations. When multiple WBANs coexist within each other's interference range, some sensors in the overlap region transmit using the same frequency to undesired receivers that could cause co-channel interference.

There is very little or no redundancy in the data measured, as the position of sensors is fixed according to its desired functionality. Thus, it is very critical to transmit the data accurately from all the sensors. If there is an overlap between the two or more WBANs, the received SINR at the LC will drop below a certain threshold. Lower value of SINR indicates strong interference and it becomes necessary to schedule the transmission of such sensors in order to avoid any packet loss. This information is broadcast by LC to its end devices along with beacon signals. The LCs which are in each other's communication range will hear this schedule and save it in its table and a minimum weight matching is created based on SINR values. This schedule is advertised just before the start of next beacon period. The sensors which suffered more interference are given priority to transmit in the modified TDMA transmission schedule. The interfering WBANs can be visualized as a bipartite graph where one set contains set of all sensor ids from the interfering WBANs and other set contains set of available transmission slots. No two sensor nodes are allowed to match with the same time slot in the same transmission round as it will cause collision. A perfect minimum match is obtained when an interfering sensor id with minimum SINR is mapped to first available time slot. Thus, a perfect match is desirable in order to avoid packet drop. Thus, we propose a solution of finding an appropriate transmission slot for interfering sensor node by using a weighted bipartite graph, where weights on the edges indicate SINR received by LC. The objective function of this optimization problem is to minimize inter-WBAN interference by assigning sensors with lower SINR to first available slot.

Thus, for such optimization, we associate a cost matrix $(c_{\mathrm{ij}})_{\mathrm{NxN}}$ by assigning a weight $c_i(j)$ to each edge connecting a sensor node i to jth transmission slot based on SINR at LC. The resulting binary matching matrix for a perfect match is given by $(m_{\mathrm{ij}})_{\mathrm{nxn}}$ where $m_{\mathrm{ij}} = 1$ if and only if the sensor node i transmits in time slot j during that round; otherwise $m_{\mathrm{ij}} = 0$ [7].

Let p_i be the probability that the transmission of node i is successful in time slot j for round number K. The expected number of successful transmissions in the next round can be given by Eq. 10.1 [8], where D_i is the number of slots passed after the node i transmitted and $K(i)$ indicates that the node i transmits in Kth round.

$$E = \sum_{i=1}^{N} pi(Di) + K(i) \tag{10.1}$$

Thus, we propose to achieve effective scheduling by obtaining assignment with an ideal match scheduling and reduce the packet drop rate. Therefore, instead of fixed TDMA with many unused data slots, we employ our matching algorithm to increase the system throughput. We use rate of packet drop per second as a performance evaluation parameter. As a part of ongoing work, we will also incorporate delay due to rescheduling in evaluating the network performance.

10.4 Performance Evaluation

10.4.1 Algorithm Implementation

Following steps are performed in order to suppress inter-WBAN interference.

- Each LC generates TDMA transmission schedule for its end devices based on the priority of the network traffic [2].
- LCs constantly computes received SINR from the nodes in its network.
- When two or more WBAN overlaps, interference causes drop in SINR. Drop below certain threshold, say 0 db is unacceptable [3], as the signal is undistinguishable from interference and noise.
- TDMA Schedules of interfering nodes along with degraded SINR advertised to neighbor LCs.
- A matching of bipartite graph containing nodes with minimum SINR values and available transmission slot based on highest to lowest interference expected by all node created using Hungarian algorithm [9].
- A modified schedule shared among LCs which assigns unique slots to interfering nodes.

10.4.2 Simulation Setup

We used Shox network simulator [10] to set up the multiple WBAN coexistence scenario with 22 sensor nodes placed by defining positions of individual sensor nodes in a 60 $\times$ 60 m area, and 3 LCs, sensor nodes are used for monitoring purposes. The properties of Physical and MAC layer are set according to IEEE 802.15.6 standards. The nodes movement is decided using Random Waypoint model with a minimum speed of 1 m/s. A variable disc model is used in which the receiver receives packet with a signal strength of $s(rx) = s(tx)/d\hat{A}^2$, where d is the Euclidean distance between a sender and a receiver pair) [10]. An Added Noise Mangler is used as bit mangling model, where a packet is dropped if the SINR falls below 0 db. LC stores the SINR values in interference queue along with the sensor ids. The lowest SINR sensor gets first priority to transmit, where as other interfering sensors are delayed until next available time slot and so on until all sensors transmit in a superframe. Hungarian algorithm is implemented in order to find perfect matching between interfering nodes and transmission slots. The modified TDMA schedule queue is broadcast to all LCs so that each one has the collision free transmission schedule.

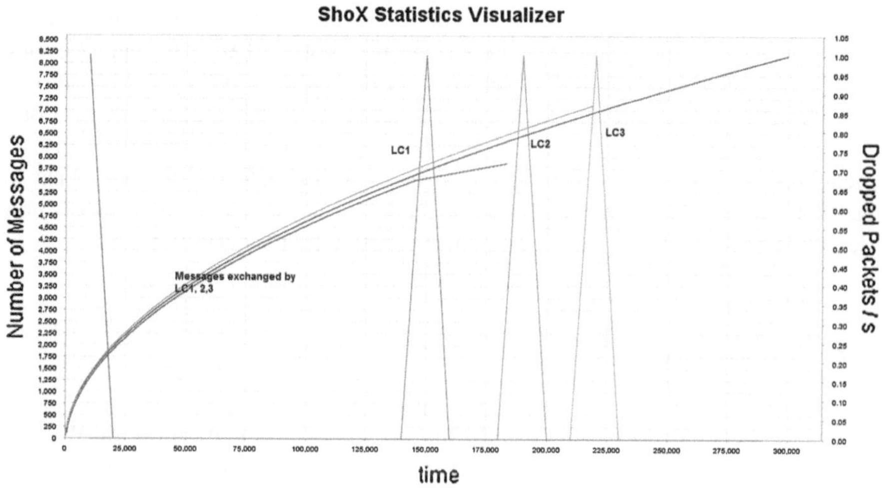

Fig. 10.3 Packet drop due to interference

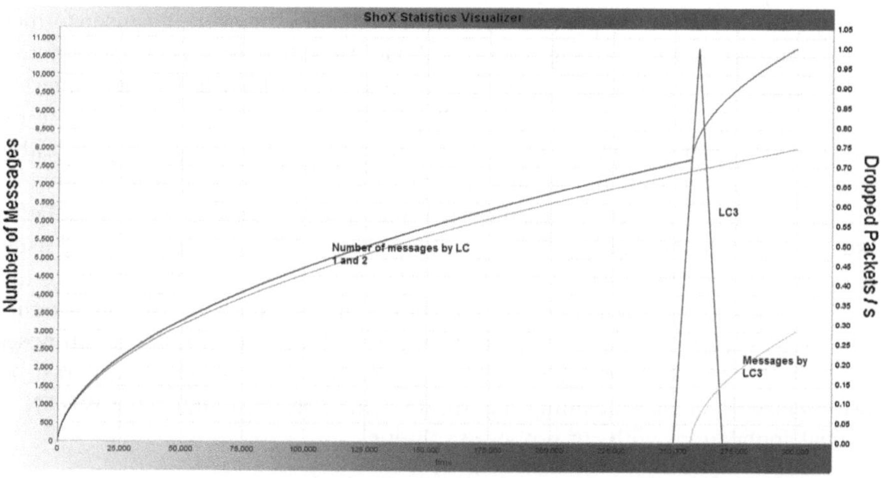

Fig. 10.4 Interference suppression using minimum weight match

10.4.3 Results and Discussion

We first simulated the scenario of 22 nodes and 3 LCs without making any change in the scheduling scheme proposed and obtained the results as shown in Fig. 10.3.

Figure 10.3 shows packet drop/s during in 3 LCs due to interference. On right side of Y axis is Dropped packet per second, which is indicated by triangular wave, whereas on left side of Y axis is number of messages exchanged, indicated by curve. X axis represents time in milliseconds.

We then repeated the simulation by adding the interference mitigation technique discussed in the above section in the Application layer, MAC layer and Physical layer for the same scenario and observed the results shown in Fig. 10.4. Although we were not able to completely eliminate packet drop, interference observed at only LC3. We were able to mitigate interference in 2 LCs. Thus, we can justify that the scheduling using minimum weight match helps to alleviate interference. Also, in Fig. 10.4, the number of messages exchanged between the interfering nodes is considerably high so as to mitigate interference as shown by the curve in figure.

10.5 Conclusion

WBAN has become very popular in healthcare applications. As multiple WBANs co-exist in a large area, there is a high possibility of inter-WBANs interference. It is expected that in 2014 there will be approximately 420 million units of wireless devices active. Thus, we will have to face the challenge of graceful coexistence of WBANs. Loss of critical data can prove to be life threatening in medical applications. An appropriate scheduling algorithm is needed which can mitigate interference. In this paper we proposed a solution of mapping interfering sensor to their transmission slots and creating a modified TDMA schedule, based on minimum SINR match. The SINR received by LCs from sensors which suffer more interference is minimum amongst all and thus they should be given higher priority in the transmission schedule. Thus, a Hungarian minimum weighted matching algorithm implemented in the simulations shows the performance improvement of network. The network performance packet drop rate per second metric is used. Simulation results are very encouraging and suggest that interference can be suppressed using minimum weight matching scheduling technique. We are currently trying to add more parameters for performance evaluation and also studying the interference mitigation when 60 sensors are present in 6 m $\times$ 6 m $\times$ 6 m space as proposed in IEEE 802.15.6 standards [10]. We would also like to compare and contrast the performance of other scheduling algorithms which are best suited for WBAN in medical applications with our proposed solution.

References

1. Jamthe, A., Chakraborty, S., Ghosh, S.K., Agrawal, D.P.: An implementation of wireless sensor network in monitoring of Parkinson's patients using received signal strength indicator. DCOSS, pp. 442–447 (2013)
2. IEEE 802.15 task group 6. (Online). Available: http://www.ieee802.org/15/pub/TG6.html and http://standards.ieee.org/findstds/standard/802.15.6-2012.html
3. Tseilshchev, Y., Libman, L., Boulis, A.: Reducing transmission losses in body area networks using variable TDMA scheduling. In: IEEE International Symposium on a World of Wireless, Mobile and Multimedia Networks, pp. 1–10. Lucca, Italy (2011)

4. Mahapatro, J., Misra, S., Manjunatha, M., Islam, N.: Interference mitigation between WBAN equipped patients. In: Ninth International Conference on Wireless and Optical Communications Networks (WOCN), pp. 1–5. (2012)
5. Prabh, K., Hauer, J-H.: Opportunistic packet scheduling in body area networks. In: Proceedings of European Conference on Wireless Sensor. Networks, Germany (2011)
6. Cheng, S.H., Huang, C.Y.: Coloring-based inter-WBAN scheduling for mobile wireless body area networks. IEEE Trans. Parallel Distrib. Syst. **24**, 1 (2013)
7. Easley, D., Klienberg, J.: Networks, crowds and markets: reasoning about a highly connected world. (Online). Available http://www.cs.cornell.edu/home/kleinber/networks-book/networks-book.pdf
8. Tseilshchev, Y., Boulis, A., Libman, L.: Variable scheduling to mitigate channel losses in energy efficient body area networks. Sensors **12**, 14692–14710 (2012)
9. Goemans, M.: Lecture notes on bipartite partitioning, Combinatorial Optimization, Massachusetts Institute of Technology, (2009). (Online). Available: http://pdfsdb.com/pdf/1-lecture-notes-on-bipartite-matching-mit-mathematics-8309080.html
10. Shox network simulator. (Online). Available: http://shox.sourceforge.net/

Chapter 11
A Cluster-Based Coordination and Communication Framework Using GA for WSANs

Arun Kumar and Virender Ranga

Abstract Wireless Sensor and Actor Networks (WSANs) are made up of a large number of sensors, small number of actor nodes, and there might be one or more base station(s) depending on the application requirement. The sensor nodes are autonomously small devices with several constraints like battery backup, computation capacity, communication range, and storage, while actor nodes are much better capable than sensors. Sensors are equipped with transceivers to gather information from their vicinity and pass it to a certain base station through actor node(s), where the measured parameters can be stored and made available for the end user. Therefore, the main issue is to send information faster and reliably with less energy consumption to the receiver node so that appropriate decision can be taken accordingly. In this paper, a new framework based on genetic algorithm (GA) is discussed with multi-tier clustering technique to transmit the data to the sink node using those actor node(s) that have more caching capability without retransmission of lost packets. The simulation results confirm the effectiveness of proposed framework over traditional approach.

Keywords WSANs • Clustering • Genetic algorithm • HEED

11.1 Introduction

Wireless Sensor and Actor Networks (WSANs) are extension to the wireless sensor networks (WSN) which includes both actor and sensor nodes. Actors are resourceful devices that can make decisions and can coordinate with each other

A. Kumar (✉) · V. Ranga
Department of Computer Engineering, National Institute of Technology Kurukshetra, Haryana, India
e-mail: arun.mahlan@gmail.com

V. Ranga
e-mail: virender.ranga@nitkkr.ac.in

R. Maringanti et al. (eds.), *Proceedings of Ninth International Conference on Wireless Communication and Sensor Networks*, Lecture Notes in Electrical Engineering 299, DOI: 10.1007/978-81-322-1823-4_11, © Springer India 2014

to perform action(s) on the basis of information reported by sensor node(s). One of the examples of actors is robots. Robots/actors can move, communicate, coordinate with other actor(s), and can take decisions, and have capability to perform action(s). On the other side, sensor node(s) sense the environment and transmit the sensed data to the sink. Sensors are small in size; low cost with limited energy, computation capability, and transmission powers. More precisely, sensor nodes are passive while actors are active in nature having more energy power, more computation capabilities, and better transmission power. However, actors and actuators are used interchangeably in the literature, but there is a minor difference in both. Actors are nodes, which take decisions but could not physically perform the action(s) while actuators can do so. Moreover, actors have the capability to coordinate with each other and take decision collaboratively to assign task to one of the actor nodes. They can work as cluster head (CH), sink, data collector or gateway depending on the network architecture employed. Their architectures could be automated, semi-automated, single-actor, or multi-actor architecture as per application requirement.

11.2 Literature Review

A number of clustering techniques have been proposed in the literature [1–16]. Most of the algorithms like LCA, RCC, CLUBS, EEHC, LEACH, FLOC, ACE, HEED, DWEHC, MOCA, and GS^3 generate variable cluster count while some algorithms have fixed cluster count. CH is selected randomly by most of the algorithms, while GS^3 approach uses preassigned CH from the sensor nodes, and the network lifetime depends on the life of CH. The network shuts down as soon as CH depletes its energy. The authors have used mobile base station to minimize the energy consumption, where energy of only CH was taken into consideration.

Hybrid Energy Efficient Distributed Clustering (HEED) [3] is a distributed clustering scheme in which CH nodes are picked from the deployed sensors based on their residual energy and communication cost when selecting CHs. HEED has three main characteristics: (1) The probability of being CH, which will be in transmission range of each other is very less. (2) Energy consumption is not assumed to be uniform for all the nodes. (3) For a given sensor's transmission range, the probability of CH selection can be adjusted to ensure inter-CH connectivity.

Energy Efficient Hierarchical Clustering (EEHC) is given by Bandyopadhyay et al. [4], which maximizes the network lifetime. It is a randomized distributed clustering algorithm for WSNs. CHs collect information from sensor nodes and perform aggregation before transmitting this to the base station. Two stage cluster formation, i.e., initial and extended is used in this algorithm. Every sensor node broadcast message claiming itself as a CH to 1-hop neighboring nodes. The nodes that are within k-hops range of a CH receive this announcement either by direct communication or by forwarding. The sensor nodes, those are not CH and are not even competing to become CH, joins a cluster as a volunteer. If information

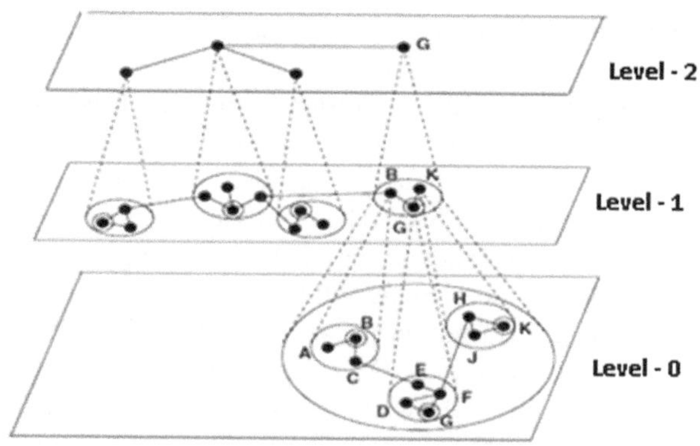

Fig. 11.1 An example of three layer cluster hierarchy [9]

regarding claim to become CH is not received, as a forwarded packet, the node will become a forced CH. Forced CH neither is a CH nor belongs to any other cluster. In the second stage, the process is applied on the CHs and the next level cluster will be chosen which will govern the underlying CHs.

Baker et al. presented Linked Cluster Algorithm (LCA) [5] in which they emphasized on forming an efficient network topology that can handle with nodes mobility. Using clustering, CHs are hoped to provide direct connectivity to all nodes in the clusters and connectivity even at the time of movement. Clustering is performed on the basis of node ID and the node with highest ID is chosen in-case of tie.

Adaptive clustering [6] is given by Lin et al., in which authors proposed the idea to minimize the transmission delay in the network and distinct code is assigned to the cluster to reduce interference between clusters. The algorithm controls the cluster size by having more number of clusters but small in size.

Nagpal and Coore proposed an algorithm called CLUBS [7] that forms clusters with the conditions that none of the nodes should be uncovered. Each cluster must have same size diameter and nodes should be able to talk with other cluster nodes. Each node in the network takes part in the cluster formation and clusters are formed including nodes with a maximum of two-hop distance. The node forwards a message with a number that decreases on every hop travel and when countdown stops, it sends a recruit messages to all nodes. Neighbors, on receiving recruit message, within two-hop boundary will accept the invitation and joins the cluster. Member nodes of cluster are termed as follower and can be part of more than one cluster.

Random competition-based clustering (RCC) [8] is designed for mobile ad hoc networks. It is also applicable to WSNs. RCC mainly strives for the cluster stability and it is based on the first come first serve strategy. The node that first asks to be CH becomes the CH itself to govern others. To remain CH, the current CH need to periodically broadcast the packets to its neighbor nodes informing its occupancy as CH.

Hierarchical control clustering [9–11] form a multi-tier hierarchical clustering as shown in Fig. 11.1. Cluster size and the degree of overlapping are taken into consideration. Any of the nodes in the network can start clustering process. The node with lowest node ID will take precedence over other nodes and more than one node can initiate cluster formation concurrently. It is a two phase process: (1) Tree discovery and (2) Cluster formation. The tree discovery uses Breadth-First-Search (BFS) tree rooted at the initiator node.

Wang et al. [12] favors the idea of clustering that resembles the data-centric design model of WSNs. The clustering would be established by mapping a hierarchy of data attributes to the network topology. The approach is based on the well-known leader election algorithm. The base station directs the nodes to form clusters by sending information packets and the nodes on receiving these packets decides whether to become a CH or not, based on their energy level. The sensor node that wants to be the CH waits for the duration based on the remaining energy level. Nodes with more energy wait longer. Interested node broadcasts an announcement from one node to next node. During the waiting time, if a node gets a claim packet from a neighboring node, to become CH, it drops its CH bid and retransmits the received packet after incrementing the hop count field in the packet by one. Upon hearing a CH announcement from a node whose attribute is different, the recipient node establishes a new cluster for the attribute and becomes a CH.

Ding et al. [13] have proposed Distributed Weight-Based Energy Efficient Hierarchical Clustering (DWEHC) to achieve more aggressive goals than those of HEED. Basically, for generating balanced cluster size and optimizing the intra-cluster topology, DWEHC is a distributed process and has $O(1)$ time complexity. Each sensor node is responsible to calculate its weight, based on its energy level with respect to the neighboring nodes. The node having largest weight would be elected as a CH and the remaining nodes become members. The nodes directly connected are considered as first-level members. A node recursively process this type of membership in order to reach a CH with least amount of energy consumption. To limit the number of levels, every cluster is assigned a range within which member nodes lies.

Youssef et al. [14] in Multi-hop Overlapping Clustering Algorithm (MOCA), argued that guaranteeing some degree of overlap among clusters can facilitate many applications like inter-cluster routing, topology discovery and node localization and recovery from CH failure, etc. They proposed a randomized, distributed algorithm that organizes the sensors into overlapping clusters. Overlapping helps in inter-cluster routing, recovery from CH failure and topology discovery. It also ensures that each node should be either a CH or within k-hops from at least one CH, where k is a preset cluster radius.

In [15], Zhang et al. presented an algorithm called GS^3, for self-configuring a wireless network into a cellular hexagon structure and control the geographical boundary of clusters. The frequency reuse technique is used to minimize energy consumption.

In cellular hexagon structure, the area is divided into cells of equal radius R, as shown in Fig. 11.2. There are two kinds of nodes, i.e., big and small in this framework. One of the big nodes starts the clustering process and selects the CH in neighboring cells and this selected CH selects their neighbors, and so on. GS^3 can be used in both environments, i.e., static network as well as in highly dynamic network where nodes are mobile and change their location frequently.

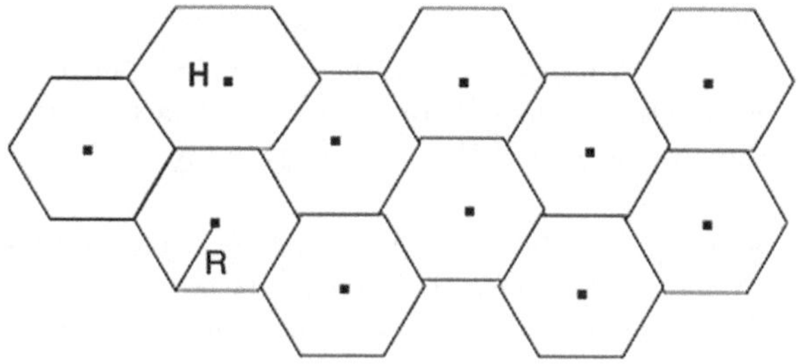

Fig. 11.2 The cellular hexagon virtual structure with big node relocated to the centre [15]

11.3 Motivation

Sensor nodes communicate with each other on a short-range channel while actor nodes can communicate over a long-range and short-range both. Whenever actors communicate with each other, they use long-range channel, but if an actor needs to communicate with sensor, it switches to short-range communication channel. The reason is that the number of actor nodes is very less as compared to the sensor nodes and actor network is prone to get partitioned whenever any or some of the actor nodes fails. Thus, to utilize actor and sensor nodes in better way, we can use the communication power of both, i.e., short and long range, to transmit data over the network. Since sensor nodes are large in number than actor and sensor network can be used to transmit data whenever path between actor–actor does not exists. Actor nodes may use fragmentation on data and get selective acknowledgement from the next actor in path to the sink/target actor. The actor node on the path keeps the segmented data until it receives selective/segmented acknowledgement from the next actor in path. After getting acknowledgment, it can flush the segment from the buffer data.

Therefore, our proposed approach transmits data to the target but leaving the bridging sensor node(s) about to be dead, because data may be forwarded through the same sensor node. Sensor nodes drain their energy very fast, if same sensor nodes are used as a gateway again and again. To avoid the energy drainage of the few of the sensor nodes that may be working as a gateway, the clustering can be formed on the sensor nodes. Sensor nodes will report to the nearest CH. CH and actor will be able to communicate on short-range channel for data transmission. The CH may be chosen on the basis of energy parameters. In our proposal, we use genetic algorithm (GA) to select CH both at sensor level as well as actor level.

11.4 Contribution of Our Proposal

The objective of the proposed approach is to find stable actor nodes for forwarding the data faster and reliably. More specifically, the main contributions of the proposed approach are:

- In our work, the network is silent until a connection is needed.
- Setup initial attributes of devices depend upon user defined scenario.

- Create a cluster network of these devices, and divide a bunch of devices into number of clusters based on sensor distribution. In each cluster, there is a CH, all devices (sensors) transmit there data to CH (actor), and then CH route the data to the other CH (actor) or gateway (sensor) in order to transmit data to the destination.
- Model a selection procedure, which is based upon GA for the proper selection of node head devices in the network arrangement.
- Calculate distance vector between devices, and update look up matrix with respect to distance matrix.
- Calculate path and cost with respect to source device, destination device, and lookup values between them.
- Devices start sending packets to CHs according to distance vector.
- The CH (actor) route the data gathered from devices (sensor) to other actor or sensor depends upon the distance vector between nodes in network (One can simply say it as cluster-based distance vector routing scheme).
- We also use a transport wrapper [16], in order to increase the network efficiency. The wrapper interacts with the routing protocol and obtains route information from it. Based on the route information, it establishes multiple transport sessions between consecutive actor network partitions.

11.4.1 Genetic Algorithm

The GA chooses the best candidate node that can become the CH. The inputs supplied to the GA are the energy parameters, i.e., (1) initial energy of the node, (2) residual energy of the node, and (3) average energy of the network. The wireless radio model plays a major role in energy consumption and therefore in the network lifetime too. The node is selected as CH based on the probability calculated by the GA. The GA loops for a number of generations to produce the optimum result. Every iteration is consists of the following steps: (1) selection, (2) reproduction, (3) evaluation, and (4) replacement. The fitness function checks for the solution produced in every new generation each time for optimal value. Figure 11.3 shows the flow chart of clustering process using GA in our proposed approach.

11.5 Data Transport Protocol

From previous researches [16], it has been established that end-to-end reliability obtained by using multiple phases of reliable transport results in a lower number of retransmissions as compared to the one that uses end-to-end retransmissions, and hence, is more energy efficient. Our framework allows us to exploit this result and develop a more efficient protocol based on this paradigm. We assume that

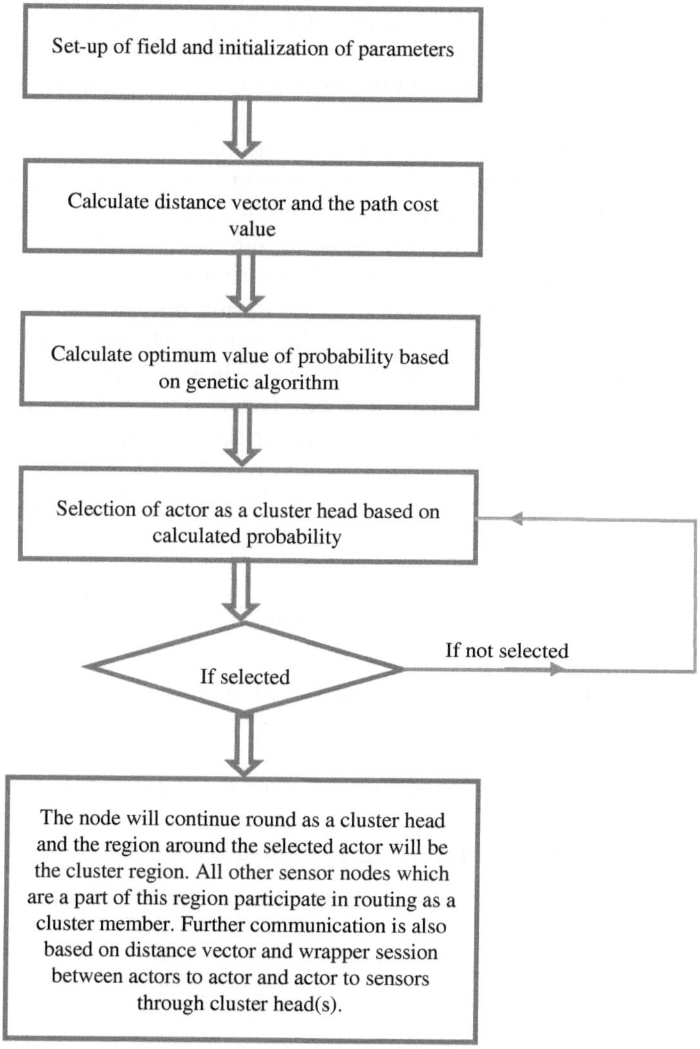

Fig. 11.3 Clustering technique based on GA

the WSAN consists of a number of partitions of actor nodes and numerous sensor nodes bridging these partitions. While sensor nodes are constrained in terms of energy and memory, actor nodes are resource rich and have large memory to store received packets and forward at a later point of time. Therefore, reliable transport between actor nodes belonging to different partitions can be obtained by setting up multiple inter-partition reliable transport sessions, one between each pair of consecutive partitions along the routes.

The data wrapper is used to transmit the data in segmented form, i.e., data is divided into segments and the actor node seeks acknowledgement for segment as well as packets transmitted. After receiving selective acknowledgment for the segment from the next actor node in the route, the actor flushes the segment from its buffer. The wrapper interacts with the routing protocol and obtains route information from it. Based on the route information, it establishes multiple transport sessions between consecutive actor network partitions. The first actor node of every partition along the route acts as the transport cache and stores the packets in the memory before it is reliably transferred to the next actor's cache along the route. The transport cache removes the packets from memory only after it receives acknowledgments from the transport cache of the next actor network partition. At the same time, there is an end-to-end wrapper session active between the source actor node and the destination actor node. The source node maintains a "master-copy" of the packets and flushes them out only when it receives an end-to-end acknowledgment from the destination node. The routing protocol informs the wrapper in case of path breaks, and the wrapper sets up a new wrapper session along the new path. Thus, the transport wrapper with sufficient cross-layer collaboration from the routing protocol achieves end-to-end reliable transport.

11.6 Pseudo Code of Proposed Approach

This section describes pseudo code of proposed approach. It has two parts: (a) Distributed Data Aggregation at sensor CH (b) Distributed Data Aggregation at actor.

Definitions: S(i): ith *Sensor node, CH(k):* kth *Cluster head, Actor(j):* jth *Actor node.*

Algorithm 1: Distributed Data Aggregation and Diffusion at Sensor CH

```
if some event e(x) detected into cluster CH(k)  then
for ∀  s(i) connected CH(k)
CH(k).Data= CH(k).Data + S(i).Data
     //[Aggregate sensed data at Cluster Head (K)]//
end for
//Perform aggregation on collected data at cluster head//
 Aggregate (CH (k))
if CH(k) not-connected to any of the Actor(j) then
Repeat (forward CH(k).Data to CH(k+1) )
 until CH(k+1) not-connected to Actor(j)
k=k+1
 else
forward CH(k).Data to Actor(j)
end if
```

Algorithm 2: Distributed Data Aggregation and Diffusion at Actor CH

```
if some event e(x) reported into the region of Actor(j) then
for ∀ CH(k) who detected event e(x) and CH(k) connected to Actor(j)
Actor(j).Data= Actor(j).Data + CH(k).Data
 end for
 end if
Repeat step 5 for all Actor nodes connected to event e(x).
if Actor(j) not-connected to Sink then
Repeat (forward Actor(j).Data to Actor(j+1))
 until Actor(j+1) not-connected to Sink
  j=j+1
 else
 for ∀ Actor(j) connected to Sink
 forward Actor(j).Data to sink
 end for
 end if
```

Actor node has more transmission power and can cover large area, so more than one CH may be connected to the one actor node. Whenever an event is detected by any of the sensor node, it immediately report to the CH_i. The CH_i can perform different aggregation function as per application's requirement(s). The CH_i after receiving sensor data from the region, it looks for the nearby actor node to forward it. Data is forwarded to the nearest actor node. So the communication takes place in between sensor nodes and the sink node through actor node(s) to speed up the transmission. Because CH may be in direct range of actor node so CH_i don't need to forward information through other sensor nodes but directly to the actor node. This will reduce the energy consumption and time delay. Even if the CH is not directly connected to the actor node then the CH_i will transmit to the next CH CH_{i-1} toward the actor node. So the end-to-end transmission delay is reduced in both the cases. So the communication takes place in between sensor nodes and the sink node through actor node(s) to speed up the transmission. Because CH may be in direct range of actor node so CH do not need to forward information through other sensor nodes but directly to the actor node. This will reduce the energy consumption and time delay. Even if the CH is not directly connected to the actor node then the CH_i will transmit to the next CH CH_{i+1} toward the actor node. So the end-to-end transmission delay is reduced in both the cases.

11.7 Performance Evaluation

In this section, we present the performance evaluation of our proposed framework using simulation experiments. Table 11.1 shows simulation parameters for evaluation.

Table 11.1 Simulation parameters

Parameters	Value
Simulation area	100 × 100
Number of sensor nodes	100, 200, 500
Number of actor nodes	20, 30, 40
Initial energy (E_0)	0.5 J/node
Transmitter electronics (E_{elec})	50 nJ/bit
Receiver electronics (E_{elec})	50 nJ/bit
Data packet size (l)	2,000 bits
Transmitter amplifier (fs)	100 pJ/bit/m^2

We consider the end-to-end delay, packet delivery ratio, network lifetime, packet loss, and throughput to evaluate the performance of proposed framework. All metrics are evaluated based on the traffic-generated from sensor node to the sink node, through actor nodes, and CH(s). We used the caching capability of actor nodes that increases throughput and consequently minimizing end-to-end delay.

The network lifetime is increased as the CH is changed periodically on the basis of energy parameters. Every time a node (sensor or actor) that have more energy level and higher probability, assigned by genetic algorithm, is selected. The clustering at sensor level and actor level decreased the end-to-end packet delay because the packets travel through the CH and CH to the sink. The CH collects the packets and transmits to the nearby actor directly.

The simulation results show the enhanced performance over the traditional WSANs. In traditional WSANs, where data retransmission and selection of the wrong CH degrades the network performance, the graphs are showing *number of actor nodes* × *number of sensor nodes* with the name of image. For example 20 × 100 means number of actors = 20, number of sensor nodes = 100. The following parameters are used to evaluate the performance of our proposed framework.

11.7.1 End-to-End Delay

This metric depicts the delay observed from sender to receiver. The end-to-end delay is decreased as the number of rounds increases, because the numbers of the intermediate nodes which are used to forward the data store stores the packets received from the previous actor/sender node until delivered to next actor in path or the original receiver. This reduces end-to-end delay in our proposed approach as confirmed in Fig. 11.4.

Because the transport protocol proposed here uses the buffers of the intermediate actor nodes while transmission. Therefore, the original sender node transmits the segment to the next actor node in path to the receiver.

The intermediate actor stores the packets received from the previous actor/ sender node until delivered to next actor in path or the original receiver. When acknowledgement for the segment is received the intermediate actor node deletes

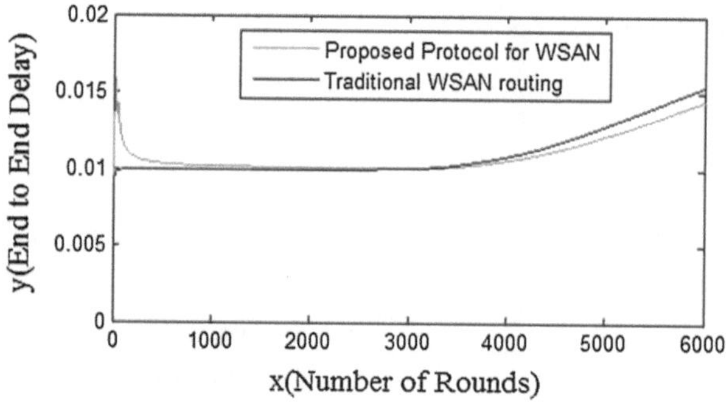

Fig. 11.4 End-to-end delay (20 × 100)

Fig. 11.5 Network lifetime

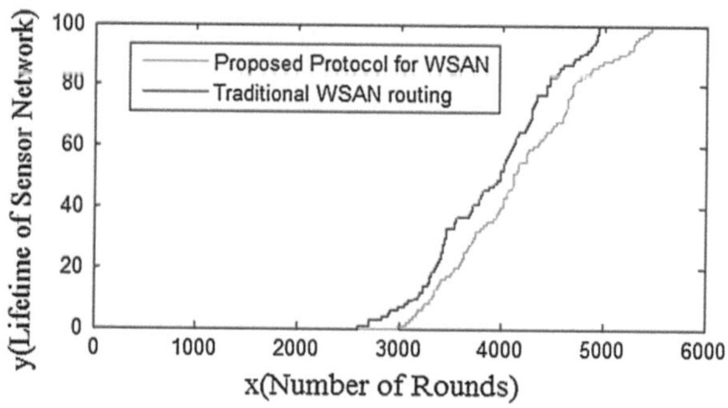

the segment. Therefore, the packet delivery performance will remain high in our proposed approach as shown in Fig. 11.5.

11.7.2 Network Lifetime

This metric is used for observing lifetime of network. A node can be dead due to some physical damage or might be out of battery power due to battery exhaustion. A network is reliable if the node death rate is low, i.e., number of alive nodes is more. A reliable network will have a better data gathering rate, i.e., data received at base station will also be high. Figure 11.6 illustrates the number of rounds for last node death as function of number of nodes in the network. Proposed approach outperforms due to use of stable nodes as CHs to transmit data through multi-hop paths instead of direct transmission to BS. Moreover, as the network size increases, the probability to find eligible stable nodes is more which is confirmed in Fig. 11.6. The GA also helps in choosing the best CH on the basis of residual energy of the node.

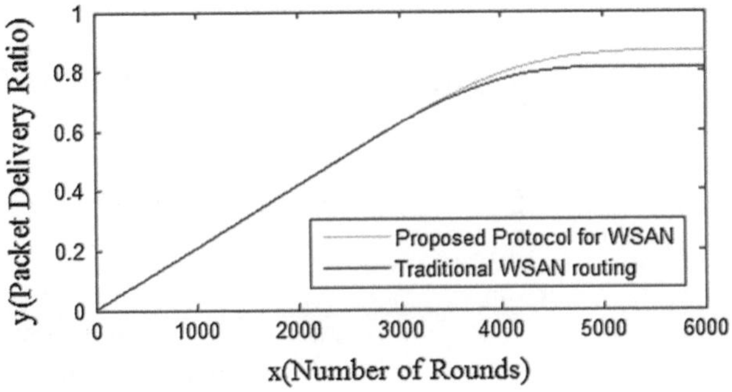

Fig. 11.6 Packet delivery ratio (20 × 100)

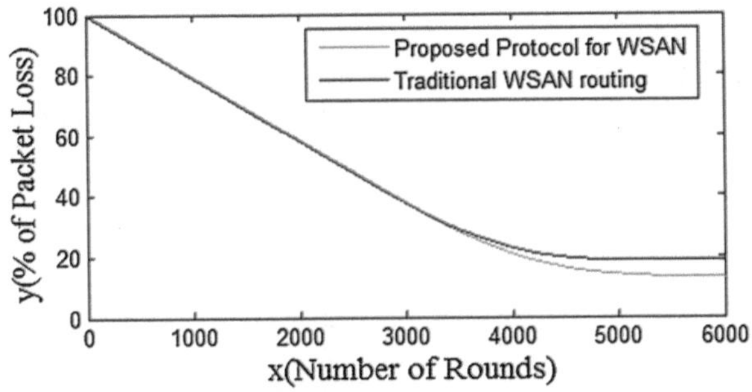

Fig. 11.7 Packet loss (20 × 100)

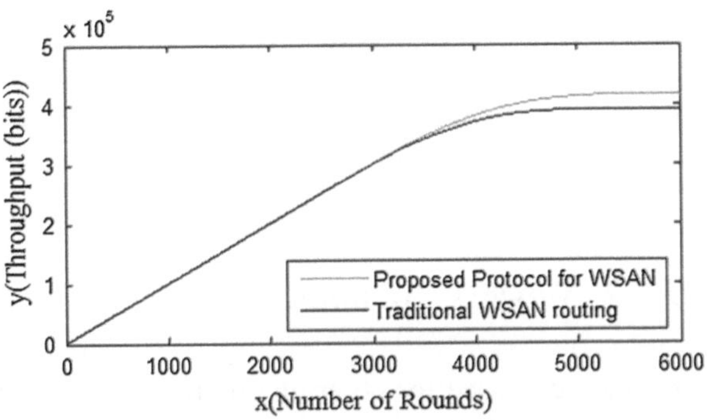

Fig. 11.8 Throughput

11.7.3 Packet Loss

Figure 11.7 depicts the packet loss in the network as the function of number of actor and sensor nodes in the network. It is observed from the simulation graph that proposed approach observes less packet loss as compared to traditional approach. The reason is that proposed approach uses intermediate acknowledgment packets during data transmission without using flooding mechanism.

11.7.4 Throughput

Due to the storage capability of the actor nodes the throughput increases and the direct communication between sensor's CHs and actor nodes also makes it faster. The sensors have limited storage capability and they can forward data to the actor at fast speed and actor is capable to handle more than one cluster data in its memory. Figure 11.8 shows that advantage of our proposed approach over traditional routing protocol.

11.8 Conclusion and Future Scope

The transportation of data is the most energy consuming process when number of packet failure increases in the wireless sensor network. To minimize packet failure, we have used reliable transport protocol that transmits the data in segmented form to the nearest actor node. This intermediate actor node saves the segment until next node does not send acknowledgment for the complete segment. It avoids end-to-end retransmission, but intermediate node, i.e., immediate packet forwarder that has stored the segment, provides the lost message immediately. The framework presented in this paper improves not only end-to-end transmission delay but also improves throughput and packet delivery ratio with the help of the multi-tier clustering. In the future, our plan is to include intermittence node(s) failure to further improve network reliability.

References

1. Loscri, V., Morabito, G., Marano, S.: A two-level hierarchy for low-energy adaptive clustering hierarchy (TL-LEACH). Vehicular Technology Conference, vol. 3, pp. 1809–1813, Sep (2005)
2. Lindsey, S., Raghavendra, C.S.: PEGASIS: power-efficient gathering in sensor information systems. In: IEEE Aerospace Conference Proceedings, vol. 3, pp. 1125–1130, (2002)
3. Younis, O., Fahmy, S.: HEED: a hybrid, energy-efficient, distributed clustering approach for ad hoc sensor networks. IEEE Trans. Mobile Comput. **3**(4), 366–379 (2004)
4. Bandyopadhyay, S.,Coyle, E.: An energy efficient hierarchical clustering algorithm for wireless sensor networks. In: Proceedings of the 22nd Annual Joint Conference of the IEEE Computer and Communications Societies (INFOCOM 2003), San Francisco, California, Apr 2003
5. Baker, D.J., Ephremides, A.: The architectural organization of a mobile radio network via a distributed algorithm. IEEE Trans. Commun. **29**(11), 1694–1701 (1981)
6. Lin, C.R., Gerla, M.: Adaptive clustering for mobile wireless networks. IEEE J. Sel. Areas Commun. **15**(7), 1265–1272 (1997)
7. Nagpal, R., Coore, D.: An algorithm for group formation in an amorphous computer. In: Proceedings of the 10th International Conference on Parallel and Distributed Systems (PDCS'98), Las Vegas, NV, Oct 1998
8. Xu, K., Gerla, M.: A heterogeneous routing protocol based on a new stable clustering scheme. In: Proceeding of IEEE Military Communications Conference (MILCOM 2002), Anaheim, CA, Oct 2002

9. Banerjee, S., Khuller, S.: A clustering scheme for hierarchical control in multi-hop wireless networks. In: Proceedings of 20th Joint Conference of the IEEE Computer and Communications Societies (INFO-COM'01), Anchorage, AK, Apr 2001
10. Demirbas, M., Arora, A., Mittal, V.: FLOC: a fast local clustering service for wireless sensor networks. In: Proceedings of Workshop on Dependability Issues in Wireless Ad Hoc Networks and Sensor Networks (DIWANS'04), Palazzo dei Congressi, Florence, Italy, June 2004
11. Chan, H., Perrig, A.: ACE: an emergent algorithm for highly uniform cluster formation. In: Proceedings of the 1st European Workshop on Sensor Networks (EWSN), Berlin, Germany, Jan 2004
12. Wang, K., Ayyash, S.A., Little, T.D.C., Basu, P.: Attribute-based clustering for information dissemination in wireless sensor networks. In: Proceeding of 2nd Annual IEEE Communications Society Conference on Sensor and Ad Hoc Communications and Networks (SECON'05), Santa Clara, CA, Sept 2005
13. Ding, P., Holliday, J., Celik, A.: Distributed energy efficient hierarchical clustering for wireless sensor networks. In: Proceedings of the IEEE International Conference on Distributed Computing in Sensor Systems (DCOSS'05), Marina Del Rey, CA, June 2005
14. Youssef, A., Younis, M., Youssef, M., Agrawala, A.: Distributed formation of overlapping multi-hop clusters in wireless sensor networks. In: Proceedings of the 49th Annual IEEE Global Communication Conference (Globecom'06), San Francisco, CA, Nov 2006
15. Zhang, H., Arora, A.: GS^3: scalable self-configuration and self-healing in wireless networks. In: Proceedings of the 21st ACM Symposium on Principles of Distributed Computing (PODC 2002), Monterey, CA, Jul 2002
16. Handigol, N, Selvaradjou, K, Murthy, C.S.R.: A reliable data transport protocol for partitioned actors in wireless sensor and actor networks. High Performance Computing (HiPC), 2010 International Conference on, pp. 1–8, 19–22 Dec 2010

Chapter 12
An ACO-Based Efficient Stagnation Avoidance Methodology for MANETS

Vinaykumar M. Kolli and G. S. Sharvani

Abstract The efficiency and behavior of a MANET depends on how well information can be passed around and delivered using Ant colony optimization (ACO). ACO algorithm heavily depends on how efficiently the pheromone is handled. Controlling pheromone is a challenging task. In this paper, an efficient pheromone decay technique is adopted which can be achieved by using stability factor 'Δ'. The stability factor is the ratio of number of Hello packets received to the number of Hello packets sent. This dynamic changing ratio will help to decide the extent of pheromone to be decayed. The technique adopted is also termed as altered exponential decay technique (AET).

Keywords ACO • Ad hoc network • Stagnation • Stability factor • Pheromone • Decay

12.1 Introduction

Ant colony optimization (ACO) algorithms have shown progress for developing routing algorithms for ad hoc networks. ACO-based routing is an efficient routing scheme based on the behavior of ants. The collective behavior of ants help to find the shortest path from the nest to a food source, by deposition of a chemical substance called pheromone on the visited nodes [1–2]. ACO was introduced in

V. M. Kolli (✉) · G. S. Sharvani
Department of Computer Science and Engineering, R. V. College of Engineering, Bangalore 560098, India
e-mail: vmkolli@yahoo.com

G. S. Sharvani
e-mail: sharvanigs@rvce.edu.in

R. Maringanti et al. (eds.), *Proceedings of Ninth International Conference on Wireless Communication and Sensor Networks*, Lecture Notes in Electrical Engineering 299,
DOI: 10.1007/978-81-322-1823-4_12,

1996, which was inspired by an algorithm called "ant system (AS)" [3]. An ant system chooses the best path laid by the previous ant, which went in search of food and has returned. ACO deals with artificial systems, which are inspired from food foraging behavior of real ants, which can find optimal solutions despite changes in the environment. The main idea is indirect communication (stigmergy) between the ants by means of pheromone trails, which helps them to find shortest path between their nest and food. Stigmergy is a major concept of ACO in which the traces in the form of pheromone left by other ants act as foundation for future coordination.

It was observed that all the social insects cooperate in an organized manner, however, looking at an individual insect they were self-directed and not being involved in the collective activity. A major problem with ACO algorithm is "stagnation" [4] or premature convergence to local optimum. This occurs when all ants try to follow same path to reach the destination (since there is more pheromone). This when applied in MANETs comes to a convergence state (equilibrium) and attracts all the data packets to follow the same path, which leads to congestion. The next packet without any awareness of the congestion follows a nonoptimal path and loses its packets due to frequent packet drops.

12.1.1 Techniques to Improve Stagnation in ACO

Stagnation is mitigated using different methodologies in ACO. In Privileged Pheromone Laying approach, selected subsets of ants are used to update pheromone values on the best path. This reduces the probability of ants following the stagnant paths that are nonoptimal and congested due to overload. In the Pheromone–Heuristic Control, ants not only try to find best path based on pheromone concentration on that edge, but also on other factors like queue length, delay, and distance. These factors alter the selection of the best path and avoid the stagnation. Pheromone Control approach is based on controlling the pheromone concentration based on adaptive nature of the network.

Pheromone control can be done in many ways. These approaches discourage nonoptimal paths and reduce the influence from past experience. In Evaporation approach all the edges with certain pheromone concentration evaporates as the time increases. This is done by an evaporation factor called "p." Evaporation not only removes the stale entries, it also balances pheromone concentration in optimal paths. This helps other ants to survey new paths (better). Aging is another technique to reduce stale entries in the network. This approach is based on the fact that older ants deposit lesser pheromone as compared to younger ants. Older ants are those ants that have taken longer time to reach the destination. Both Aging and Evaporation aims at finding new best path when there is congestion. Pheromone can be limited for every edge by placing an upper bound. A variant of the Limiting pheromone is pheromone smoothing in this approach the pheromone is increased along an edge. It is also observed that smaller amount of pheromone is deposited gradually until the upper bound is reached. Evaporation is done in a uniform manner on all edges. This technique seems to be more effective in avoiding the generation of dominant path [5–7]. The similarity between Social insects like ants and routing in AWN using agents makes ACO an efficient approach for routing in AWN. The paper attempts to analyze ACO-based routing algorithm that have been developed for AWN. The survey aims at creating awareness for the research scholars about the techniques used in ACO to avoid stagnation, highlight its methodology, etc.

12.2 Design of Algorithm

The probability at source node V_i can be computed based on the packet forwarding equation. This depends on Pheromone, Distance, and Packet delivery ratio of the path. The integral solution for the above combination of equation decides the roots i, j which extracts the path to be followed or the path in which packet has to forwarded. Z_i is the probability that the node(i) will get selected by the current node for packet forwarding as shown in Eq. (12.1). Once Z_i for all i's where i represents the neighbor of the current node are calculated, Maximum among the $Z[i]$ are calculated and node corresponding to the Maximum $Z[i]$ is determined as the node to which packet has to be forwarded.

$$Z_i = \frac{\mathrm{K} * \frac{\mathrm{Pheromone}(i) * \mathrm{PDR}(i)}{\mathrm{Distance}(i)}}{\sum_{i=1}^{n} \frac{\mathrm{Pheromone}(i) * \mathrm{PDR}(i)}{\mathrm{Distance}(i)}} \quad (12.1)$$

$$\forall i \in S, \text{ Set of Nodes}$$

Max($Z(i)$) where i will be the node to which packet will get forwarded. K is the constant to balance the equation which do not impact much on Z_i. The pheromone value for each node is initialized to 0.015. During Routing using Ant Colony Optimization, the pheromone content of the node involved in the routing is increased irrespective of whether the node is source or destination or any intermediate node during packet transmission. Periodically, the pheromone content is decreased as per the equation. For all $i \in Nn$, For all $d \in Dn$, The pheromone between I and D at nth Second can be given as shown in Eq. (12.2).

$$\mathrm{Pn}_{i,d} = \mathrm{Pn}_{i,d} * e^{-\tau+\Delta} \quad (12.2)$$

For all $i \in Nn$, For all $d \in Dn$, the pheromone between I and D at nth Second. The decay rate is $\tau \geq 0$, and is a global parameter. To account the successful packet transmission, the pheromone on all the links present in the path traversed are updated. Presently, the pheromone update is taken as 0.05 for each link on the path traversed. The links are virtually classified into two categories viz., stable links and unstable links.

The extent of decrement is determined by time difference between the current time and last successful transaction (Rank of Freshness of the link) and also the stability factor, which is generic to any link. Its value is decided after running through several models of analytics and routing scenarios.

In this paper, an exploration on approximation of pheromone value in the exponential decay can be controlled using additive value which helps in controlling the trail evaporation. Various values for τ has been tested and came to the conclusion that the following values forms the most fitted curve for the decay rate. Hence, they are taken as the testing parameters under decay rate. 'Δ' is the stability factor which acts as pheromone controlling agent for exponential decay.

12.3 Results and Discussions

Different Decay rate values for τ associated with the stability factor, collectively deciding the extent of pheromone decay are tested using network simulator software developed as part of this work. These values are tested under various possible real-time scenarios usually faced in Mobile Ad hoc Networks such as mobility pattern of nodes, velocity, and packet delay, etc. The ACO without pheromone decay technique (PDT) will experience congestion whereas the ACO with PDT will avoid the congestion due to efficient decay (controlled manner) of pheromone on all the links.

12.3.1 Decay Rate (τ) Used in PDT Versus PDR (Data Goodput)

Figure 12.1 shows the obtained results for different values of decay rate and it shows that the value 0.001 is best suited in most of the scenarios. Also from the figure we can observe that the decay rate is acceptable for the range of values 0.001–0.003 in most of the situations. The values of 0.01 and 0.1 are applicable in very rare scenarios of MANETs.

12.3.2 Probability of Path Selection Using Pheromone Concept

The period of decay, at the end of tick of which the pheromone will get decremented by specific amount irrespective of the conditions or the future packets about to be sent which are likely to follow the same path. The decay rates decide the extent of decrement and also periodicity is the factor which accounts to final PDR of the particular scenario. The same analysis has been made on a MANET scenario using simulator developed. The results of the same are as shown in Fig. 12.2. The decay rate value 0.001 shows better performance in terms of higher probabilities of path selection.

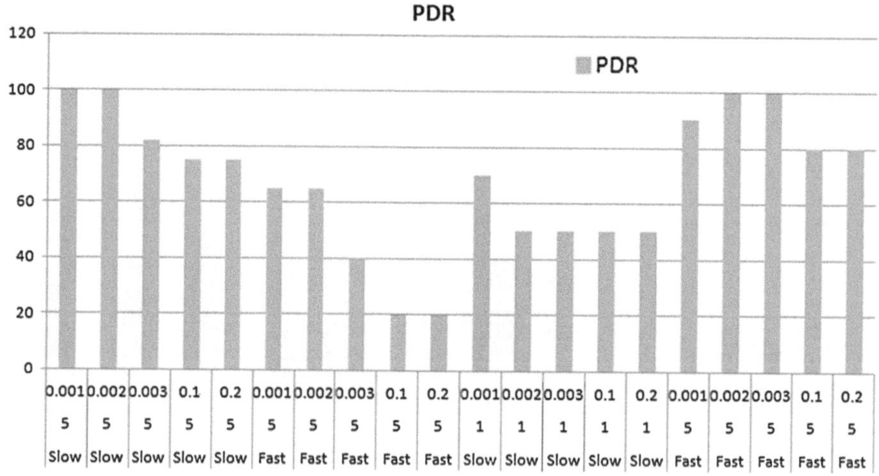

Fig. 12.1 Combined view of decay rates versus PDR

Fig. 12.2 Probability of path selection for various decay rate

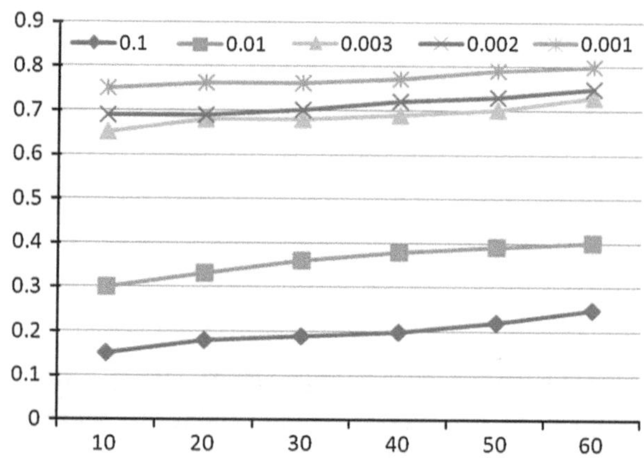

12.3.3 Effect of Stability Factor in Routing

The stability factor represents the local status of a node in a network and accounts the same during calculating the extent of decay due to which the pheromone decay value becomes dynamic and adaptable to the networking scenarios, in particular the pheromone control adapts to the nodes mobility, the packet transmission table that it has, and also the past history of it about the successful transmissions went through some specific reliable neighbor.

To adopt the stability factor concept into routing, new packet forwarding equation is prototyped and implemented. The impact of stability factor may be negligible in cases where distribution of nodes is highly dense and nodes are moving

Fig. 12.3 Pheromone intensity with time plotted against the decay period adopted

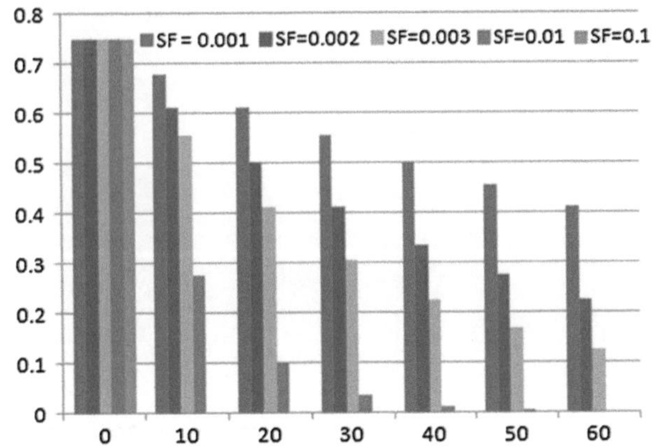

Fig. 12.4 Number of nodes versus PDR

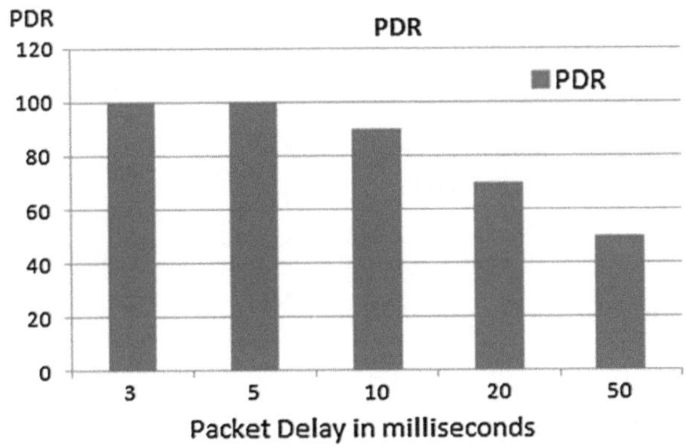

relatively. But in cases where QoS and security are given importance, stability factor plays the major role in deciding the reliable path thereby giving the Best Quality of Service (QoS).

12.3.4 Number of Nodes Versus PDR (Data Goodput) and PACKET Delay of Nodes Versus PDR

As number of nodes increases, density of distribution of nodes in the network area increases; this increases the PDR of network scenarios. Figure 12.3 shows the obtained results. If the packet delay is less, the subsequent packet transmissions efficiently utilizes the pheromone laid on the network links by the previous successful transmissions.

The link change rate measures the average number of links that change state, new or leaving, every second at each node. It is a relative measure of how fast the network topology is changing, and thus the time available to acquire new information about the network. This metric is shown in Fig. 12.4 for each tested speed and node count.

It is also found that, as the speed of the node increases, the link change rate will increase rapidly.

12.4 Conclusion

After the thorough analysis of proposed technique following inferences can be drawn. ACO Routing with Pheromone Decay Technique is proven to be **more efficient** than that without pheromone decay. The Goodput obtained in the case of PDT is always more than or equal to the one without PDT. ACO Routing with **PDT never suffer from Congestion** whereas the ACO without PDT will definitely encounter congestion depending on the routing scenarios. The goodput not only depends on the pheromone concentration but also **on different network metrics** such as Time Duration of Routing, Speed of Nodes, Mobility Method used, etc. The Goodput also depends on the **number of nodes in the network**, however, it cannot be considered as the performance metric in the present context. Stability factor determines local reliability thereby accounting to good overall throughput and more suitable to adopt only in the cases wherein the QoS and Security is more important. Decay rate of 0.001 with stability factor factor $\cong$100 % in dense distributed randomly moving pattern of ad hoc nodes is considered to be the best value for the decay for controlling the stagnation as well as allowing the routing to use the advantages of pheromone-based ACO methodologies. The scenario mentioned in the above inference is the most generally observed environment in ad hoc networks. This inference is obtained by analyzing the bulk of real-time network topology with different requirements.

Performance of each accounting method scales with network size although there is a general negative trend. The improved performance is gained by moving away from the biological perspective and borrowing ideas from traditional linear filtering. The pheromone sensitivity and decay rate also have an effect on performance. There seems to be a sensitivity threshold above which performance is unchanged, while the pheromone decay rate should be more carefully determined. Their relative effects can offset each other to some extent as well. Results are generally very good, delivering the large majority of packets to their destination without any control traffic under very high mobility conditions.

References

1. Chandra Mohan, B., Baskaran, R.: A survey: Ant colony optimization based recent research and implementation several engineering domain. Int. J. Expert Syst. Appl. (Elsevier Publications), 4618–4627 (2012)
2. Ducatelle, F., Caro, G.A.D., Gambardella, L.M.: Principles and applications of swarm intelligence for adaptive routing in telecommunications networks. In: Proceedings of Swarm Intelligence, pp. 173–198 (2010)

3. Dorigo, M., Gambardella, L.M., Birattari, M., Martinoli, A., Poli, R., Stützle, T.: ACO and swarm intelligence. In: 5th International Workshop, ANTS 2006. LNCS 4150. Springer, Berlin (2006)
4. Sim, K.M., Sun, W.H.: Multiple ant colony optimization for load balancing. In: IDEAL, pp. 467–471 (2003)
5. De Rango, F., Tropea, M., Provato, A., Sanmaria, A.F., Marano, S.: Minimum hop count and load balancing metrics based on ant behavior over hap mesh. In: IEEE GLOBECOM, pp. 1–6, New Orleans (2008)
6. De Rango, F., Tropea, M.: Energy saving and load balancing in wireless adhoc networks through ant based routing. In: SPECTS, vol. 41 (2009). doi: 978-1-2-4244-4165-5
7. Cowdhury, N.M., Syed, M., Choudhury, E.H.: A new adaptive routing approach based on ACO for ad hoc wireless networks. In: Proceedings of International Workshop on Internet and Distributed Computing Systems, pp. 51–56 (2008)

Chapter 13
An Approach to Network Coding at Data Link Layer

Vivekanand Jha, Nidhi Nagpal, Anchal Goswami and Bhavnit Kaur

Abstract Majorly, the technique of store-and-forward is employed in networked systems for transmitting information through a network. However, the optimality of this approach was challenged by network coding theory. The network information flow problem has proved to be solved by network coding, as it optimizes the flow of data in a network in an effective way. Network coding can improve the throughput, robustness, and security of a network and has already been implemented at the physical layer, network layer, and is proposed at transport layer. In this paper, we have proposed a complete binary tree approach based on virtual channels for implementing network coding at data link layer. The simulation results have shown reduction in delay, number of transmissions, congestion in the network, and better utilization of bandwidth; thus, it has been shown that this approach led to significant improvements in the performance of a network by effectively utilizing the single channels as in case of link layer.

Keywords Network coding • Encoding • Decoding • Complete binary tree (CBT) approach • Virtual channels

V. Jha (✉) · N. Nagpal · A. Goswami · B. Kaur
Department of Computer Science and Engineering, Indira Gandhi Delhi Technical University for Women, Delhi, India
e-mail: vivekanand.iiitm@gmail.com
URL: http://www.igit.ac.in

N. Nagpal
e-mail: nidhinagpal91@gmail.com

A. Goswami
e-mail: anchal.igit@gmail.com

B. Kaur
e-mail: kaurbhavnit@gmail.com

R. Maringanti et al. (eds.), *Proceedings of Ninth International Conference on Wireless Communication and Sensor Networks*, Lecture Notes in Electrical Engineering 299,
DOI: 10.1007/978-81-322-1823-4_13,

13.1 Introduction

In this paper, the preliminaries of Network Coding, its variants and techniques and the functionality of link layer are discussed in brief.

13.1.1 Traditional Routing Versus Network Coding

In this section, we compare traditional routing and network coding. In traditional routing, data received at a node is simply selectively retransmitted by the node. In other words, routing in a network uses store-and-forward strategy. Network coding is an extension of routing which allows for any node in the network to perform operations on its received data before it transmits any data [1]. Hence, it is a method of attaining maximum information flow in a network. It is an elegant technique introduced to improve network performance and throughput. With network coding, a node is allowed to combine a number of received packets and then create one or several outgoing packets from them. So, outgoing packets are linear combinations of the original packets. An encoded packet generally carries information about several original packets, but in contrast to concatenation, it does not allow the recovery of original packet from the concatenated packet. Decoding information also needs to be known for recovery.

The data link layer is the second layer in the Open Systems Interconnection (OSI) seven-layer reference model. It responds to service requests from the network layer above it and issues service requests to the physical layer below it. It encapsulates network layer packets into frames and is concerned with local delivery of those frames from one hop to the next hop. This layer also includes mechanisms to detect and then correct data transmission errors. It also contains flow and error control mechanisms.

13.2 Background

Network coding has a vast research background as it has gone through various improvements and has been used in wide variety of applications. The concept of network coding was fully developed by Ahlswede et al. [1] where the term network coding was coined. Their proposed network model demonstrated the data processing at intermediate nodes. Subsequently, Li et al. [2] showed that the linear network codes sufficed to achieve the max-flow rate for the single-source multicast scenario. Christos et al. [3] introduced a new scheme for content distribution of large files that is based on network coding. Network coding was applied in single-source multicast networks till 2005. Tracey et al. [4] presented a distributed random linear network coding approach for transmission and compression of information in general multisource multicast networks. After network coding was implemented at network layer, Shenghli et al. [5] proposed network coding at physical layer. Unlike conventional network coding which performs coding arithmetic on digital bit streams after they are decoded, Physical Layer Network Coding (PNC) makes use of the additive nature of simultaneously arriving electromagnetic (EM) waves and applies the network coding arithmetic at the physical layer. Jay et al. [6] implemented network coding technique at transport layer by adding network coding specifications in TCP. Dong et al. [7] proposed some network coding schemes to reduce the number of broadcast transmissions from one sender to multiple receivers. Baochun et al. [8] conceptualized benefits of using random network coding in P2P networks. Chen et al. [9] addressed the problem of flow control at end systems for network coding-based multicast ows. Jie et al. [10] used flow-based XOR network coding for lossy wireless networks.

13.3 Motivation

Network coding already exists at network layer and physical layer and has also been proposed on the transport layer. At data link layer the communication is based on hop-to-hop delivery, if network coding is implemented at the data link layer, it can prove to be significantly beneficial. In this paper, a complete binary tree approach based on virtual channels is presented for implementing network coding at the data link layer.

13.4 Proposed Work

In the proposed solution, it is assumed that a single channel is made of several virtual channels. Each virtual channel consists of respective unique local encoding vector. Based on these local encoding vectors, each node maintains a local encoding matrix. For network coding to be optimal, number of frames available at any node must be greater than or equal to 2^k, where k is the number of hops between source and destination. If this assumption does not hold then forwarding the frames without network coding is the best possible solution. Major steps in proposed solution are discussed below:

Packet Division. As the source nodes link layer receives the packet from network layer, these are converted to frames. It is assumed that the Source S has N frames to send.

Hop Count. Since topology of network is known so from source to destination total hops are counted.

Complete Binary Tree based Virtual Channel Creation. The N frames will be linearly combined on 2^d virtual channels, where d is total hop count. And to get the maximum benefits of network coding number of incoming channels should be large as compared to number of outgoing channels. (i.e., $N > 2^d$) as shown in

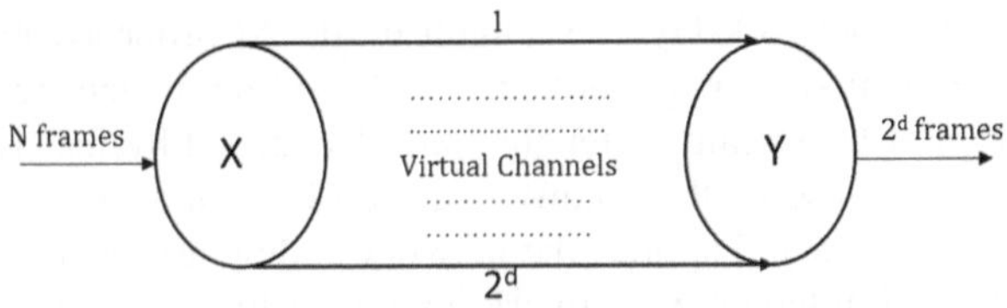

Fig. 13.1 Virtual Channel Creation

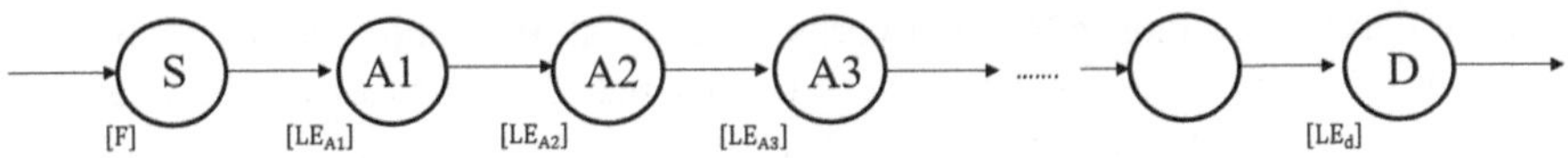

Fig. 13.2 Local Encoding Vectors

Fig. 13.1. To implement complete binary tree (CBT) structure for virtual channel next hops virtual channels number reduced by half till destination. Finally, a CBT structure is obtained from destination to source. The depth of this CBT is total hop count from source to destination.

Buffer Management at Each Hop. Each node maintains two buffers, which store Encoded Frames: an Input Buffer whose size is equal to the number of incoming virtual channels and an output buffer whose size is equal to the number of outgoing virtual channels. Sender and destination nodes keep only one buffer. Sender node uses its buffer to store the encoded frames and destination node stores the two received encoded frames in its buffer.

Encoding and Decoding. Each node consists of local encoding vector (LE) and input and output buffers which are used to combine linearly all input frames. The incoming encoded frames are stored in the Input Buffer and once encoded by the node using local encoding vectors; these are stored in the output buffer before being sent over the channel along various output virtual channels. The local encoding vectors are LE_{A1}, LE_{A2}, …, LE_d as shown in Fig. 13.2.

The orders of the matrices are shown below:

$$
\begin{aligned}
&[F] : N \times 1 \\
&[LE_{A1}] : 2^d \times N \\
&[LE_{A2}] : 2^{d-1} \times 2^d \\
&[LE_{A3}] : 2^{d-2} \times 2^{d-1} \\
&\quad\vdots \\
&[LE_d] : N \times 2 \\
&[X] : N \times 1
\end{aligned}
$$

Here, $[LE]$ stands for local encoding vector local encoding vectors for respective virtual channels at link S-A_1 are as follows:

$$[LA_1] = \left[g_1^1, g_2^1, \ldots, g_N^1\right]_{1\times N} \tag{13.1}$$

$$[LA_2] = \left[g_1^2, g_2^2, \ldots, g_N^2\right]_{1\times N} \tag{13.2}$$

$$\vdots \tag{13.3}$$

$$\left[LA_2^d\right] = \left[g_1^{2^d}, g_2^{2^d}, \ldots, g_N^{2^d}\right]_{1\times N} \tag{13.4}$$

These vectors of each virtual channel combined together give the local encoding matrix stored at the source node shown in Eq. (13.5).

$$[LE_{A1}] = \begin{pmatrix} g_1^1 & g_2^1 & \cdots & g_N^1 \\ g_1^2 & g_2^2 & \cdots & g_N^2 \\ \vdots & \vdots & \ddots & \vdots \\ g_1^{2^d} & g_2^{2^d} & \cdots & g_N^{2^d} \end{pmatrix} \tag{13.5}$$

This matrix, when multiplied by the matrix $[F]$ which is the incoming frames vector of order $N \times 1$, gives the matrix $[Y_{A1}]$ of resultant order $2^d * 1$. Similarly, we have $[LE]_{A2}$ and corresponding $[Y_{A2}]_{2^{d-1}x1}$, and so on. And $[LE]_D$ and $[Y_D]2X1$. Global encoding vector, G is calculated my multiplying all the local encoding vectors in the following manner as shown in Eq. (13.6).

$$[G]_{N\times N} = [LE_D]_{N\times 2} \ldots [LE_{A3}]_{2^{D-2}X2^{D-1}} \times [LE_{A2}]_{2^{D-1}\times 2^D} \times [LE_{A1}]_{2^D\times N} \tag{13.6}$$

Encoding process is multiplication of matrices in following Eq. (13.7):

$$[X] = [F] \times [G] \tag{13.7}$$

where, $[G]$ is global encoding vector. This is known to destination for decoding purpose. The decoding process is done at the destination using the following Eq. (13.8).

$$[X] = [F] \times [LE_{A1}] \times [LE_{A2}] \times [LE_{A3}] \times \ldots [LE_d] \tag{13.8}$$

where

$[F]_{N\times 1} = [G]_{N\times N-1} \times [X_D]_{N\times 1}$
$[G]_{N\times N-1}$: Inverse of the Global Encoding Vector
$[X_d]_{N\times 1}$: Received encoded frames at the destination.

13.5 Error Control Mechanism

For the proper delivery and acknowledgement of encoded frames, we use selective repeat ARQ, one of the error control mechanism available at data link layer. The selective repeat ARQ is deployed on the respective buffers containing encoded

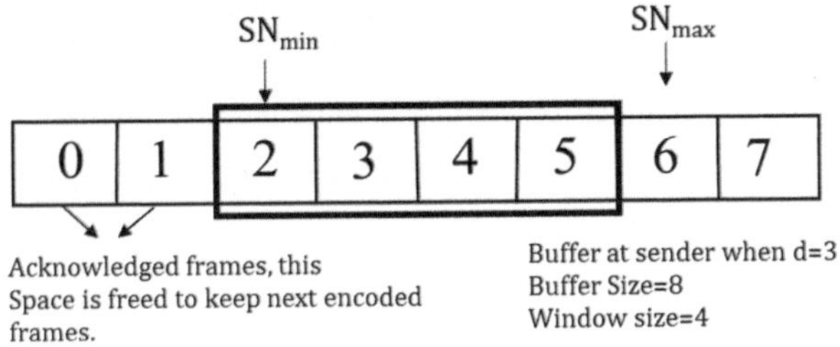

Fig. 13.3 Window Management

frames, not on the received frames. The sequence numbers of encoded frames are chosen using modulo-m operation as shown in Eq. (13.9)

$$m \geq 2 \times \frac{\text{Buffer Size}}{2} \tag{13.9}$$

(as for selective repeat, $m \geq [2 \times \text{window size}]$)

Window Size, n, is calculated using Eq. (13.10)

$$n = \frac{\text{Buffer Size}}{2} \tag{13.10}$$

As described in Fig. 13.3, window slides when acknowledgement is received for a frame. Frames with sequence number 0 to ($SN_{\min} - 1$) are deleted from buffer and this memory is freed to keep next encoded frames. Window slides when acknowledgement is received for a frame.

$SN_{\min}$smallest numbered frame that has not yet been acknowledged.
$SN_{\max}$the number of next frame to be accepted from the higher layer.

13.6 Minimization of Delay

The number of hops between the source and the destination must not be very high for network coding to be optimal. If this happens to be high then the amount of delay is very high as the source node waits for large number of frames before encoding and forwarding. If this assumption does not hold true, then the proposed solution needs to be modified such that the delay introduced in minimum. If there are a large number of hops between the sender and the destination, i.e., the value of d is high then we divide the linear network, having d hops between the Source and the Destination, into subnetworks of k hops each (except the last subnet, which can have $\leq k$ hops). As shown in Fig. 13.4, the source and the destination of all the sub-networks are temporary sources and destinations called codecs (except the Source of the 1st subnet and Destination of the last subnet) and network coding or simple forwarding can now be applied within each subnet. A Codec refers to an intermediate node that acts as a temporary destination as well as temporary source at times. It decodes the received frames and again encodes them and sends to the next hop.

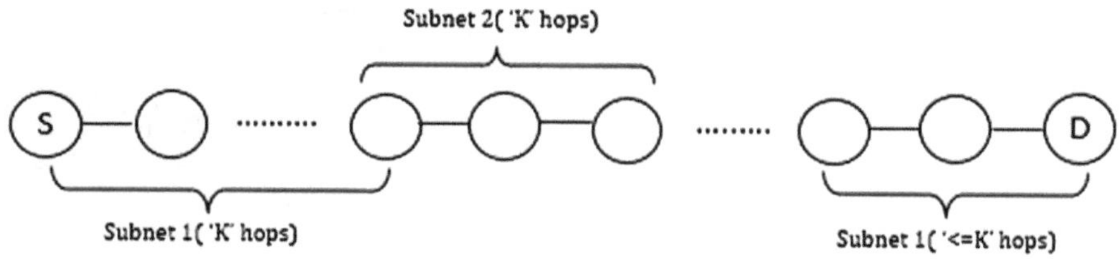

Fig. 13.4 Minimization of Delay

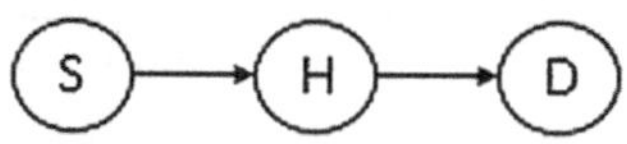

Fig. 13.5 The implementation topology in Netbeans IDE

13.7 Implementation Details

The Network coding at data link layer has been implemented in two parts. First, the proposed work has been implemented on Netbeans IDE and second, the readings for performance measures are taken by simulating network coding at data link layer on the simulator NS2.

Netbeans IDE The Network Coding at data link layer has been implemented on Netbeans IDE for the topology shown in Fig. 13.5. The Source, *S* reads four characters input from in file, converts them into a frame and forwards these frames to the hop one after the other. The Hop has a buffer of size four which stores the incoming frames. Once this buffer is full, the hop encodes these four frames into two frames by multiplying these with the local encoding vector. These two frames are forwarded to the destination, *D*, one after the other. The Destination then decodes these two frames to recover the original for frames by calculating the global encoding vector.

Simulator NS2 The topology for computing the performance measures is depicted in Fig. 13.6. Node 1 and Node 2 send one frame each to Node 3 via channel 1 and channel 2, respectively. Node 3 encodes the two received packets and sends the encoded packet to node 4. Node 4 then decodes two original packets from the encoded one. All the links have a bandwidth of 0.3 Mb/s, and a propagation delay of 10 ms. The buffer size on the links is set at 200. The TCP receive window size is set at 8 packets and the packet size is 52 bytes.

13.8 Simulation Results and Analysis

Time taken for different transmissions in this whole communication has been noted to evaluate certain parameters, and thus respective improvements for evaluating performance parameters in NS2. The Table 13.1 shows the sending and receiving time of packets at different nodes without using network coding.

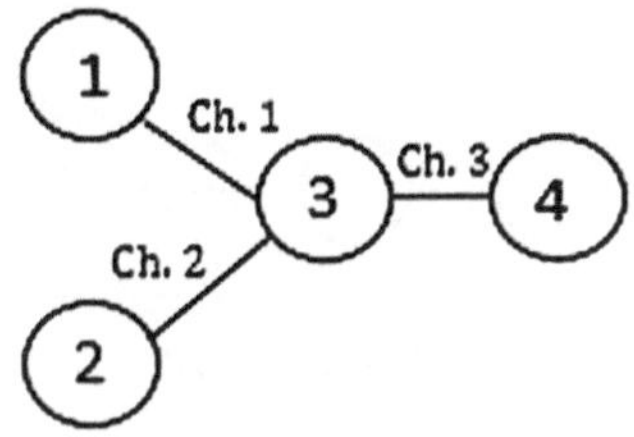

Fig. 13.6 The NS2 topology

Table 13.1 Without network coding

Packet no.	Sending time	Receiving time
1	0.130	0.238
2	0.130	0.238
3	0.240	0.430
4	0.256	0.451

Table 13.2 With network coding

Packet no.	Sending time	Receiving time
1	0.130	0.238
2	0.130	0.238
3	0.331	0.436

The Table 13.2 shows the sending and receiving time of packets at different nodes using network coding.

We present the results with the help of graphs drawn using the values of time from Tables 13.1 and 13.2. As shown by the graph of Fig. 13.7, time taken by packet 1, 2 from nodes 1, 2, respectively, to reach next hop, i.e., node 3 is same. Without using network coding technique, node 3 forwards both the packets one by one, while when network coding is used, one transmission of encoded packets works in the place of two transmissions, hence taking lesser amount of time to reach node 4, the destination. In this particular case, Network coding reduces total communication time by 5.6.

Figure 13.8 shows the variation of number of packets in various channels with respect to time. When network coding is used, at time 0.434 s, channel 3 contains encoded packet and therefore it is the last transmission via this link. While without using network coding, channel 3 contains one more packet at time 0.451 s. Hence, it shows that network coding reduces the traffic at channel and also takes lesser time. The graph of Fig. 13.9 shows the values of throughput in different channels using network coding. The size of each packet is 52 Mb. The formula shown below has been used to calculate the throughput. Throughput = (Packet size)/(time taken to send one packet) As shown by the graph for channel 3, when network coding is used throughput improves by a good amount, since it takes lesser time and communicates two packets. The three packet transmissions that are used in this network communicate the contents of two packets. Without coding, these three transmissions cannot be used to communicate as much information and they

Fig. 13.7 Graph for Time vs. Packet Number

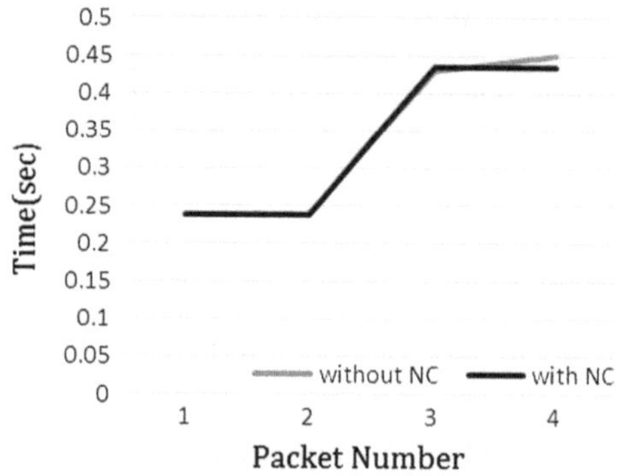

Fig. 13.8 Variation of number of packets in various channels with respect to time

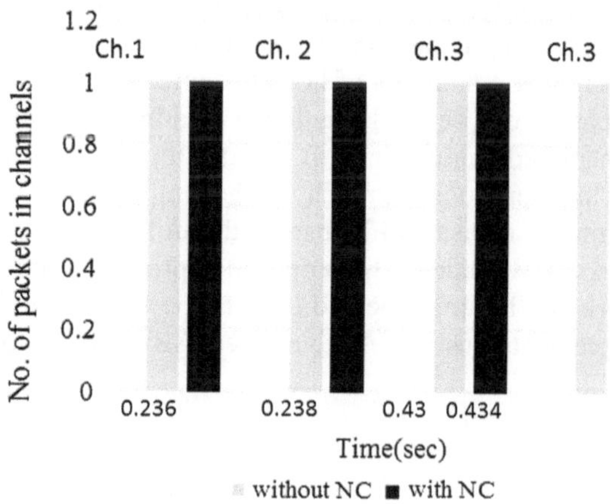

Fig. 13.9 Values of throughput in different channels using network coding

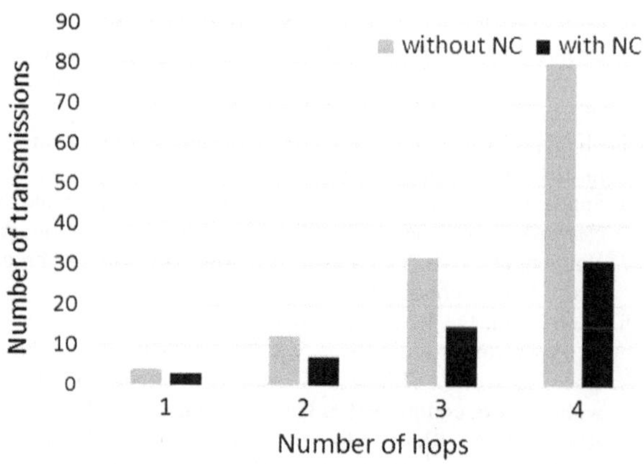

must be supplemented with additional transmissions. Figure 13.10 shows the variation of total number of transmissions with number of hops. This plot assumes:

Fig. 13.10 Variation of total number of transmissions with number of hops

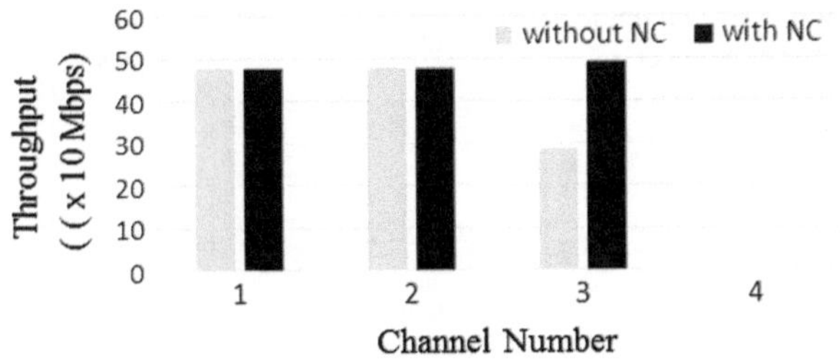

$N = 2^d$ where N is the number of frames sent by source, and d is the number of hops between source and destination.

In this paper, we have presented a complete binary tree approach to implement network coding at data link layer. We have also proposed a technique to minimize the delay when numbers of hops become very large. It is thus concluded that network coding can be implemented at data link layer and this will reduce the number of transmitted frames, lead to a better throughput and minimize the delay in network. In future, our approach can be simulated for a bigger topology. Some other performance measures like bandwidth utilization, congestion control, and security can also be evaluated. The proposed approach for minimizing delay in a very large network can be implemented on a frame-based simulator. This approach can also be implemented using selective repeat ARQ and Go-back n protocol.

References

1. Ahlswede, R., Cai, N., Li, S.Y.R., Yeung, R.W.: Network information flow. IEEE Trans. **46**(4), 1204–1216 (2000)
2. Li, S.-Y.R., Yeung, R.W., Cai, N.: Linear network coding. IEEE Transl. Inf. Theor. (2003)
3. Gkantsidis, C., Rodriguez, P.R.: Network coding for large scale content distribution. IEEE (2005)
4. Ho, T., Leong, B., Koetter, R., Medard, M.: Distributed asynchronous algorithms for multicast network coding (2006)
5. Zhang, S., Liew, S.-C., Lam, P.P.: On the synchronization of physical-layer network coding. In: IEEE Information Theory Workshop (2006)
6. Sundararajan, J.K., Shah, D., Medard, M., Jakubczak, S., Mitzenmacher, M., Barros, J.: Network coding meets TCP: theory and implementation. Proc. IEEE **99**(3) (2011)
7. Nguyen, D., Tran, T., Nguyen, T., Bose, B.: Wireless broadcast using network coding. IEEE Trans. Veh. Technol. 58(2) (2009)
8. Li, B., Niu, D.: Random network coding in peer-to-peer networks: from theory to practice. In: Proceedings of the IEEE, vol. 0018–9219 IEEE 99, No. 3 (2011)
9. Chen, L., Ho, T., Chiang, M., Low, S.H., Doyle, J.C.: Congestion control for multicast flows with network coding. IEEE Trans. Inf. Theor. **58**(9) (2012)
10. Khreishah, A., Khalil, I.M., Ostovari, P., Wu, J.: Flow-based XOR network coding for lossy wireless networks. IEEE Trans. Wireless Commun. **11**(6) (2012)
11. Bhatia, J., Patel, A., Narmawala, Z.: Review on variants of network coding in wireless ad-hoc networks. IEEE (2011)
12. Hu, Y., Yu, C.-M., Li, Y.K., Lee, P.P.C., Lui, J.C.S.: NCFS: On the practicality and extensibility of a network-coding-based distributed file system. IEEE (2011)

13. Iqbal, M.A., Dai, B., Huang, B., Hassan, A., Yu, S.: Survey of network coding-aware routing protocols in wireless networks. J. Netw. Comput. Appl. **34**, 1956–1970 (2011)
14. Zhang, G., Zhang, G., Cheng, S.: LANC: locality-aware network coding for better P2P traffic localization. Comput. Netw. **55**, 1242–1256 (2011)
15. Yang, Z., Li, M., Lou, W.: R-code: network coding-based reliable broadcast in wireless mesh net-works. Ad Hoc Netw. **9**, 788–798 (2011)
16. Fan, Y., Jiang, Y., Zhu, H., Chen, J., Shen, X.S.: Network coding based privacy preservation against traffic analysis in multi-hop wireless networks. IEEE Trans. Wireless Commun. **10**(3) (2011)
17. Fragouli, C., Le Boudec, J.Y., Widmer, J.: Network coding: an instant primer. ACM SIGCOMM Comput. Commun. Rev. **36**(1) (2006)
18. Bhatia, J., Patel, A., Narmawala, Z.: Review on variants of network coding in wireless ad-hoc networks. In: International Conference on Current Trends in Technology, NUiCONE (2011)

Chapter 14
Event Triggered Multipath Routing in Wireless Sensor Networks

A. V. Sutagundar, S. S. Manvi and N. C. Debnath

Abstract Reliable communication, optimal usage of energy and bandwidth are the critical issues in wireless sensor networks (WSNs). Efficient utilization of WSN resources prolongs the network life time. In this paper, we propose an event triggered multipath routing algorithm for WSNs. The proposed scheme operates in the following steps. (1) The sensor node (first sensed) that detects an event (called as event node) triggers multiple path discovery from itself to the sink node. (2) Event node initiates the route initialization phase where it sends the beacon packet to neighbor sensor nodes. The beacon packet comprises of event type (critical/non-critical), node id, location information, residual energy, available bandwidth, and hop distance. (3) Sink node computes the node disjoint paths. (4) Sink node computes the path weight factor based on the energy efficiency, path efficiency, and bandwidth efficiency. (5) If event is noncritical, then sink node selects a path with highest weight factor (based on cost function) and sends the path information to event node. For critical information, multipath information is sent to event node. To test effectiveness of the proposed scheme, it is analyzed in terms of packet delivery ratio, energy consumption, overhead for critical and noncritical information.

Keywords Wireless sensor network • Multipath routing • Event triggered

A. V. Sutagundar (✉)
Department of Electronics and Communication, Basaveshwar Engineering College, Bagalkot, Karnataka, India
e-mail: sutagundar@gmail.com

S. S. Manvi
Department of Electronics and Communication, REVA Institute of Technology and Management, Bangalore, India
e-mail: sunil.manvi@revainstitution.org

N. C. Debnath
Winona State University, Winona, MN 55987, USA
e-mail: Ndebnath@winona.edu

R. Maringanti et al. (eds.), *Proceedings of Ninth International Conference on Wireless Communication and Sensor Networks*, Lecture Notes in Electrical Engineering 299, DOI: 10.1007/978-81-322-1823-4_14, © Springer India 2014

14.1 Introduction

Wireless sensor networks (WSNs) comprises of a set of smart devices, called sensors, which are able to sense and transmit information about the environment on which they are deployed [1, 2]. Due to recent technological advances, the manufacturing of small and low-cost sensors become technically and economically feasible. Large number of sensors can be networked in many applications that require unattended operations like military applications, habitat monitoring, environmental monitoring, etc. These sensors have the ability to communicate either among each other or directly to an external base station (BS), called as sink. The deployment of more number of sensors allow sensing over larger geographical regions with greater accuracy. Using conventional methods of data gathering and processing in WSNs could lead to some of the problems like energy consumption, redundant data transmission, increased latency, bandwidth overheads, etc. Each sensor node has the capability to gather and route data either to other sensors or back to an external sink node. The critical issue in WSN is network lifetime which is mainly dependent on the sensor node battery. To tackle the problem, redundant transmissions can be minimized by certain techniques like data fusion, data aggregation, etc.

Routing in WSNs is very challenging due to the features that distinguish these networks from other wireless networks like mobile ad hoc networks or cellular networks [3, 4]. Due to the relatively more number of sensor nodes, it is difficult to build and maintain the global addressing scheme for the deployment of a more number of sensor nodes as the overhead of ID maintenance is high.

Some of the related works are as follows. There are several requirements for a routing algorithm in WSNs. The main objectives of WSN routing algorithm are: both energy efficient and energy balancing are to be achieved in order to prolong the lifetime of sensor networks, algorithm should follow a distributed control for scalable WSN and it needs to be robust to diverse potential event generation patterns [5]. The work given in [6] presents a work to prolong the network lifetime of WSNs using multipath routing based on a family of flexible routes with soft quality of service guarantees in terms of the packets delivery latency. The reduction in the total energy consumption of WSNs in multi-hop data aggregation is done by constructing energy efficient data aggregation trees [7]. An energy efficient clustering routing (EECR) algorithm for WSNs is presented in [11, 12]. Sensor network topology is organized into several clusters and selection of cluster head is based on the weight value and the residual energy that leads to uniform energy dissipation among the sensor nodes. A routing protocol for WSNs to support an information-fusion application is presented in [8]. In [9], agent-assisted QoS-based routing algorithm for WSNs is described. Multi-agent system is used to monitor changes in-network topology, network communication flow, and each nodes routing state. Agents can participate in in-network aggregation, routing, and network maintenance. The Nonagent-based multipath routing (NABMR) in WSNs is a hazard aware multipath reliable routing algorithm [10].

Objectives of the proposed scheme are as follows. (1) Selection of disjoint paths from event node to the sink node. (2) Computation of the weight factors for the available disjoint paths based on available residual energy, bandwidth, and distance. (3) Selection of path(s) based on the criticalness of an event.

The rest of the paper is organized as follows. Proposed event triggered multipath routing in WSNs is discussed in Sect. 14.2. Planned simulation is discussed in Sect. 14.3. Finally, Sect. 14.4 concludes the paper.

14.2 Proposed Work

In this section, we describe network environment, model for multipath routing in WSN for critical and noncritical data.

14.2.1 Network Environment

The network environment considered for data gathering, aggregation, and routing in WSNs is as shown in Fig. 14.1. It comprises of sensor nodes with diversified sensing competence and a sink node. Sensor nodes sense the data periodically and send it to sink node with multi-hop communication. In the proposed scheme an event node forms a cluster and is also called cluster head.

Some of the assumptions that are considered in this work are as follows.

- All nodes (sensors and sink nodes) in the network are static and have same initial energy.
- During deployment phase, each sensor node has full energy.
- All the sensor nodes are equipped with global positioning system (GPS), processor, and transceiver for the communication.

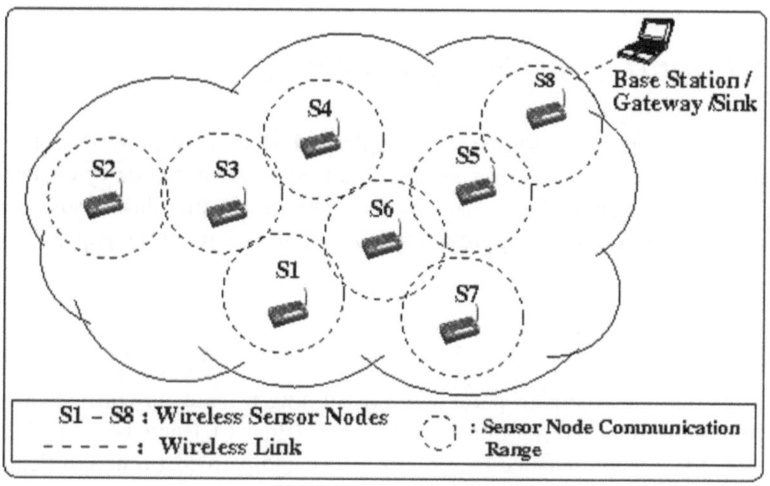

Fig. 14.1 Network environment

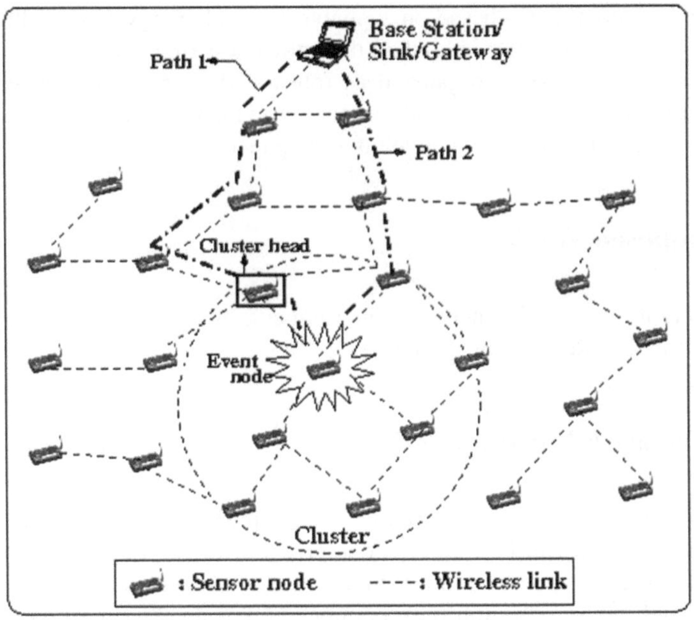

Fig. 14.2 Event triggered multipath routing in WSNs

- All nodes reconfigure their transmission power.
- A sensor node (active node) participates in aggregation if and only if the sensed values in a particular time window drifts by a given threshold.
- A sensor node which detects the event forms the dynamic cluster.
- Cluster head gathers data from all sensor nodes within the cluster.

14.2.2 Event Triggered Multipath Routing

Figure 14.2 depicts event triggered multiple path discovery in WSNs. There are two paths that are computed by sink node that lead to event node or cluster head. Cluster head aggregates all the data of sensor nodes in the cluster, and sends to sink node. The discussion of aggregation is not in the scope of the paper.

14.2.2.1 Path Parameters

Sink node computes the path parameters (assuming n nodes in path) such as path length, the minimum and maximum of the path (Eqs. 14.1 and 14.2), path energy factor (Eq. 14.3), path distance factor (Eq. 14.4), and path cost function (Eq. 14.5).

$$E_{R\min} = \mathrm{Min}\{E_R(1), E_R(2), \ldots E_R(n)\} \qquad (14.1)$$

Where $E_{R\min}$ is the minimum residual energy among the nodes in a path.

$$E_{R\max} = \mathrm{Max}\{E_R(1), E_R(2), \ldots E_R(n)\} \tag{14.2}$$

Where $E_{R\max}$ is the maximum residual energy among the nodes in a path.

Energy factor (E_{fact}) of path is given by Eq. (14.3).

$$E_{\mathrm{fact}} = \frac{E_{R\min}}{E_{R\max}}. \tag{14.3}$$

The distance factor (D_f) for each path is computed by using Eq. (14.4).

$$D_f = \frac{P_d}{P_{\mathrm{hc}}} \tag{14.4}$$

where P_d is the path distance and P_{hc} is the hop count.

Cost function, C_f of each path is given by Eq. (14.5).

$$C_f = E_{\mathrm{fact}} + D_f \tag{14.5}$$

Sink node prioritize the paths by using C_f. For noncritical information transmission, a path with highest C_f is chosen whereas for reliable critical information transmission, paths with C_f in decreasing order are selected.

14.2.2.2 Operation Sequence

The sequence of the operation for the proposed scheme is as follows. (1) Sensor nodes read the neighbor node information, which can be used for aggregation. (2) A sensor node that detects an event, initiates the route discovery process from event node to sink node. Event node also acts as cluster head, which may aggregate the information within the cluster. (3) Event node floods the beacon packet toward the sink node. The beacon packets move from node to node and reach the sink node. If a beacon packet does not find sink node within given hops, such a beacon packet expires. Duplicate beacon packets will be rejected by the nodes. (4) Beacon packets that successfully reach the sink node deliver the path information to the sink node. The information comprises of: sensed information, type of information (critical or noncritical), residual energy in the traversed nodes, distance traveled, and hop count. (5) Sink node computes the multiple paths based on information received through beacon packets using shortest path algorithm (using hop count as metric) applying iteratively by truncating selected links for the path in every iteration. Multiple paths are ranked based on the cost function as discussed above. (6) Sink node checks and verifies the criticalness of the event. Discussion of verification is not in the scope of this work. (7) A path (single path) with highest C_f is chosen for critical information, whereas for reliable critical information transmission, paths with C_f in decreasing order are selected. Sink node sends the path(s) information over the chosen first path to event node. And, (8) Event node takes the action of sending data upon receiving the path information by using aggregation, if necessary. The forwarding nodes in the path may also aggregate the information.

14.3 Simulation

The proposed scheme has been simulated in various network scenarios using C++ programming language. A discrete event simulation is done to test operation effectiveness of the scheme. In this section, we describe the simulation model and the simulation procedure.

14.3.1 Simulation Model

WSN is generated in an area of $l \times b$ square meters. It consists of N number of static nodes, placed randomly. Each node is associated with energy E_f joules and transmission range R meters. The communication environment is assumed to be contention-free. The transmission of packets is assumed to occur in discrete time. A node receives all packets heading to it during receiving interval unless the sender node is in nonactive state. We assumed the channel as error free. Sensor MAC protocol (S-MAC) [13] is used for media access. Free space propagation model is used with propagation constant β.

14.3.2 Simulation Procedure

Simulation inputs for proposed scheme are as follows: $l = 5{,}000$ m, $b = 5{,}000$ m, number nodes $N = 300$, transmitting nodes $(Nt) = 40$–200, $\beta = 2.5$, $R_c = 300$–500 m, $E_f = 1$ kJ, $T_{\text{rpkts}} = 256$ per second, $BW_{\text{single-hop}} = 5$ Mbps, size of sensed data at each node $S_d = 5$ Kbytes, size of the processing code $S_{\text{proc}} = 3$ Kbytes.

Simulation procedure for the proposed scheme is as follows.

1. Topology generation: Generate the WSN in the given area by placing nodes randomly. Each node maintains a data structure to store the information as specified by the scheme.
2. Sensor node gets the neighbor node information.
3. Apply the proposed scheme.
4. Compute performance parameters of the system.

14.3.3 Results

Figure 14.3 presents packet delivery ratio for given number of nodes involved in transmission. The amount of data generated increases with increase in number of nodes. As the number of nodes increase, the amount of data generated from source sensor nodes will increase. Hence, available bandwidth may not be sufficient for

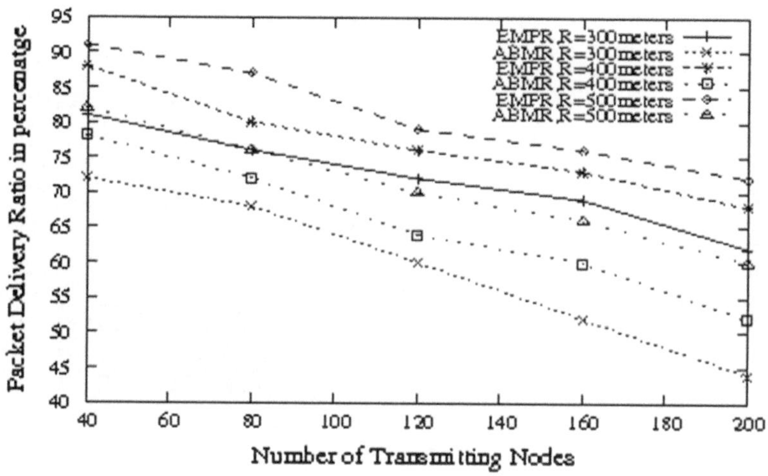

Fig. 14.3 Packet delivery ratio versus number of transmitting nodes

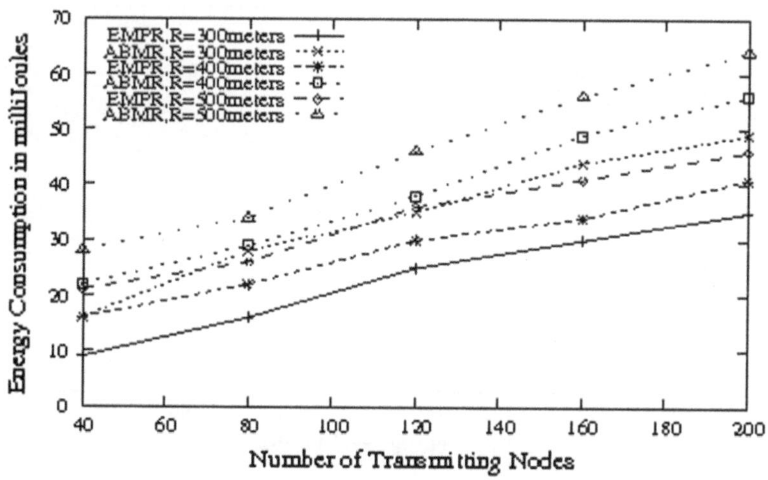

Fig. 14.4 Energy consumption in millijoules versus number of transmitting nodes

successful transmission of the data thereby causing decrease in the packet delivery ratio. We can also notice from the results that the proposed EMPR performs better compared to ABMPR.

Figure 14.4 explains the energy consumption for the given number of transmitting nodes. With increase in the number of nodes and communication range, the energy consumption increases. Energy consumption is due to gathering of partial topology information, path computation, path information transmission, and reception. The proposed scheme uses PDA for partial topology information gathering and SMA for path computation. The packet delivery ratio increases

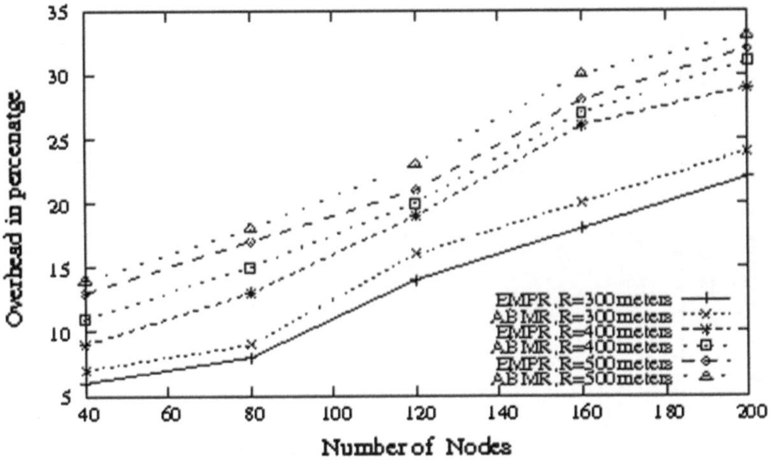

Fig. 14.5 Overhead in percentage versus number of nodes (critical)

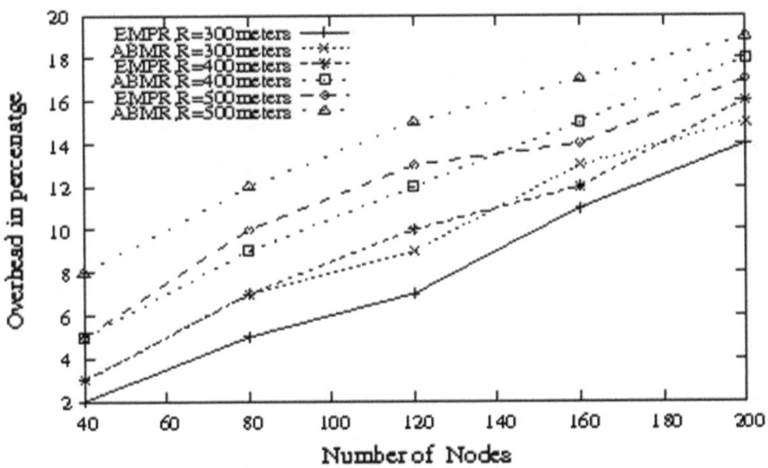

Fig. 14.6 Overhead in percentage versus number of nodes (noncritical)

with increase in transmission range since number of isolated nodes will be very less. LEDMPR performs better compared to ABMPR since LEDMPR uses partial topology information whereas ABMPR uses the total topology information.

Figures 14.5 and 14.6 presents the overhead with the given number of nodes and communication range for critical and noncritical data communication. For critical information communication, proposed scheme uses the multipath routing in order to achieve the reliable communication, whereas noncritical information communication uses the single path with highest weight factor. For increase in the number of nodes and communication range, the number of transmissions and computations will increase.

14.4 Conclusion

In this paper, we have proposed an energy efficient reliable event triggered multipath routing in WSNs. Multipaths are computed based on the path cost function depending on path energy factor and path distance factor. The proposed scheme is tested in terms of packet delivery ratio, energy consumption, overhead for critical and noncritical data. However, there are some limitations with respect to security and validation. We are planning to employ certain strategies based on certificates and digital signatures for path security and validation of beacon packets.

Acknowledgments We are thankful to Basaveshwar Engineering College, TEQIP phase-2, Karnataka, INDIA, for financial assistance.

References

1. Chong, C.Y., Kumar, S.P.: Sensor networks: evolution, opportunities, and challenges. Proc. IEEE **91**(8), 1247–1256 (2003)
2. Diallo, O., Rodrigues, J.J.P.C., Sene, M.: Real-time data management on wireless sensor networks: a survey. Elsevier J. Netw. Comput. Appl. **35**, 1013–1021 (2012)
3. Al-Karaki, J.N., Kamal, A.E.: Routing techniques in wireless sensor networks: a survey. IEEE Wireless Commun. **11**(6), 6–28 (2004)
4. Radi, M., Dezfouli, B., Bakar, K.A., Lee, M.: Multipath routing in wireless sensor networks: survey and research challenges. J. Sens. **12**, 650–685 (2012)
5. Ok, C., Lee, S., Mitra, P., Kumara, S.: Distributed routing in wireless sensor networks using energy welfare metric. Elsevier Inf. Sci. **180**, 1656–1670 (2010)
6. Ghica, O., Trajcevski, G., Scheuermann, P., Valtchanov, N., Bischof, Z.: Controlled multipath routing in sensor networks using Bezier curves. Comput. J. Adv. Access **54**, 1–25 (2010)
7. Zeydan, E., Kivanc, D., Comaniciu, C., Tureli, U.: Energy-efficient routing for correlated data in wireless sensor networks. Elsevier Ad Hoc Netw. Article in press (2012)
8. Nakamura, E.F., Ramos, H.S., Villas, L.A., de Oliveira, H.A., de Aquino, A.L., Loureiro, A.A.: A reactive role assignment for data routing in event-based wireless sensor networks. Elsevier Comput. Netw. **53**, 1980–1996 (2009)
9. Liu, M., Xu, S., Sun, S.: An agent-assisted QoS-based routing algorithm for wireless sensor networks. Elsevier J. Netw. Comput. Appl. **35**, 29–36 (2012)
10. Ramadan, R.A.: Agent based multipath routing in wireless sensor networks. In: Proceedings of IEEE Symposium on Intelligent Agents IA 09, pp. 63–69. (2009)
11. Li, L., Dong, S.S., Wen, X.M.: An energy efficient clustering routing algorithm for wireless sensor networks. Elsevier J. Chin. Univ. Posts Telecommun. **13**(3), 71–75 (2006)
12. Soro, Stanislava, Heinzelman, Wendi B.: Cluster head election techniques for coverage preservation in wireless sensor networks. Elsevier J. Ad Hoc Netw. **7**, 955–972 (2009)
13. Wendi, R., Heinzelman, A., Chandrakasan, Hari, B.: Medium access control with coordinated adaptive sleeping for wireless sensor networks. IEEE/ACM Trans. Netw. **12**(3), 493–506 (2004)

Chapter 15
Algorithm for Gunshot Detection Using Mel-Frequency Cepstrum Coefficients (MFCC)

Preetam Suman, Subhdeep Karan, Vrijendra Singh and R. Maringanti

Abstract Protection of forests and wildlife needs efficient, reliable, and real-time detection of events such as gunshots, wood cutting, distress call of animals, etc. In this paper, we propose a gunshot detection technique through acoustic signal pattern recognition utilizing Mel-Frequency Cepstrum Coefficients (MFCC). In this work, MFCC is used to extract the features of gunshots from prerecorded analog sound files. Training of the system for gunshot detection has been done using a three layer Artificial Neural Networks (ANN) using extracted parameters of acoustic signals. For the creation of the database, 150 prerecorded gunshots have been used. From the database, 80 gunshot sound samples have been used for the training of the system. Testing has been done with the remaining 70 samples in the presence of noise. The algorithm has also been tested successfully using actual gunshot in noisy environment. Efficiency of algorithm is 95 % without noise and that decreases to 85 % in the presence of noise.

Keywords Acoustics • MFCC • ANN • Gunshot detection • Association rules • Decision tree

P. Suman (✉) · S. Karan · V. Singh · R. Maringanti
Indian Institute of Information Technology, Allahabad, India
e-mail: preetam@iiita.ac.in

S. Karan
e-mail: subhadeepkaran@gmail.com

V. Singh
e-mail: vrij@iiita.ac.in

R. Maringanti
e-mail: mkrishna@iiita.ac.in

R. Maringanti et al. (eds.), *Proceedings of Ninth International Conference on Wireless Communication and Sensor Networks*, Lecture Notes in Electrical Engineering 299,
DOI: 10.1007/978-81-322-1823-4_15, © Springer India 2014

15.1 Introduction

Continuous surveillance of forests is a very difficult task for human beings. But the increased hunting, poaching, and deforestation activities make a strong case for real-time and automated surveillance system for the protection of wildlife and forests, and to identify illegal activities.

This paper presents an algorithm that was developed to detect the gunshot sound amidst forest noises and clutter. For this, 150 prerecorded gunshot sound samples [1] have been analyzed to extract the significant parameters required for the detection of gunshot (Analysis is described in Sect. 15.3). Gunshot consists of muzzle blast, shock wave, hammer mechanism during fire, and surface vibration due to muzzle blast. This paper focuses on detection of gunshot by analyzing muzzle blast sound. Feature extraction of gunshot sound has been done by Mel-Frequency Cepstrum Coefficients (MFCC) [2]. Of these 150 samples, 80 sound samples have been utilized to train a three-layer artificial neural network. The three-layers comprise of an input layer that takes sample signals, hidden layer to process input signal, compare with decision rules to identify the signal, and an output layer that produces the result [3]. After training of system, decision tree-based association rules were developed for testing of algorithm.

The Sect. 15.2 of this paper presents a short review of the literature with regard to gunshot sound characteristics, feature extraction of acoustic signal, identification, and localization of event, Sect. 15.3 presents analysis of signal using spectrogram, Sect. 15.4 presents features extraction using MFCC, Sect. 15.5 presents training of the system, Sect. 15.6 presents the association rule learning, Sect. 15.7 presents testing of the algorithm in the forest, and Sect. 15.8 concludes the paper.

15.2 Related Work

The literature is rich with a number of acoustic classification techniques.

Maher et al. [4] discussed some basic characteristics of gunshot like muzzle blast, mechanical action during firing (trigger and hammer mechanism, ejection of cartridge, etc.), supersonic projectile, and surface vibration due to muzzle blast. Each of these characteristics is useful to detect a gunshot. Author also discussed the effect of wind, humidity, temperature, and other ground obstacle on gunshot.

Chu et al. [5] presented the feature analysis for general environmental sound characterization using matching pursuit (MP) algorithm to establish time-frequency features. Authors of this paper combined the features of MP and MFCC for high accuracy in the recognition of environmental sound classification like: inside restaurants, playground, train passing, inside casinos, nature-daytime, ocean waves, raining/shower, thundering, etc. Authors of this paper implemented K-nearest neighborhood (KNN) and Gaussian mixture model (GMM) to classify the sound. Authors have collected 14 different environmental sounds to train and test the algorithm. The overall accuracy of recognition is 82.3 %, and it varied between 50 and 100 % for different environmental sounds. The most difficult environment sounds reported were ocean wave 63 %, sounds from movie 73 %, sounds inside casino 70 %, and traffic noise 74 %.

Freire [6] discussed use of correlation of the audio signal with predefined template (database of features of audio signals) to detect gunshot in a noisy environment. Three techniques linear predictive coding (LPC) coefficients, impulsivity parameter from stable distributions [7], and MFCC [2] have been used to extract the features of sound samples. The authors were able to detect the gunshot up to 30 dB SNR, and when noise is further increased to 25 dB SNR the system gives false positive result and when SNR was lowered 20 dB the system gives false negative results.

Ghiurcau et al. [8] discussed classification of different sounds originating from humans, cars, and birds to protect restricted areas like forest, lakes, natural parks, etc. The paper presents time-encoded signal processing and recognition (TESPAR) algorithm for sound classification. The Authors have created a database of 300 recorded sounds including several types of environmental noise like rain, wind, etc. The noises are recorded separately and then added to testing samples. Authors reported 94 % success to classify the sound in case of recorded sound without noise that decreases to 85 % with decreasing SNR.

Smith [9] proposed a solution for gunshot detection and localization. Location of the gun is determined by time of arrival information of muzzle blast. Time of arrival is determined by microphone embedded in JTRS radio, which acts as sensor node. Authors developed an algorithm that uses muzzle blast time of arrival information to determine the location of the shooter. The combination of correlation filters and rake receiver has been used to detect and localize gunshots. The correlation filter was used to remove uncorrelated surrounding noise from signal, and rake receiver use to eliminate multipath signals. The algorithm was tested in field with eight microphones and a GPS unit connected with laptops. The efficiency of algorithm was 90 % for actual gunshot and it has given 10 % false positive result. The testing with database (created by online available gunshots sound) was 96 %.

Aleksi [3] presented implementation of multichannel FPGA-based acoustic localization device (ALD) which can identify the direction of acoustic signal. The single board acoustic signal detector (ASD) used an electret microphone, and a signal conditioner using LM324N and a digitization unit. Eight ASDs were placed in different directions were connected to a single FPGA board through wire. The setup was tested in laboratory. A finite state machine was implemented on FPGA to recognize direction of acoustic signal. Due to simple processing, the acoustic resolution of the direction identification was not good.

15.3 Analysis of Gunshot Using Spectrogram

Spectrogram is a representation of time varying signal that shows variation of spectral density with time. The *x*-axis of spectrogram represents time, *y*-axis represents frequency, and different colors in spectrogram shows the intensity of signal. The spectral analysis of the sounds from pistol, rifle, Sniper, and shotgun are presented in the following figures.

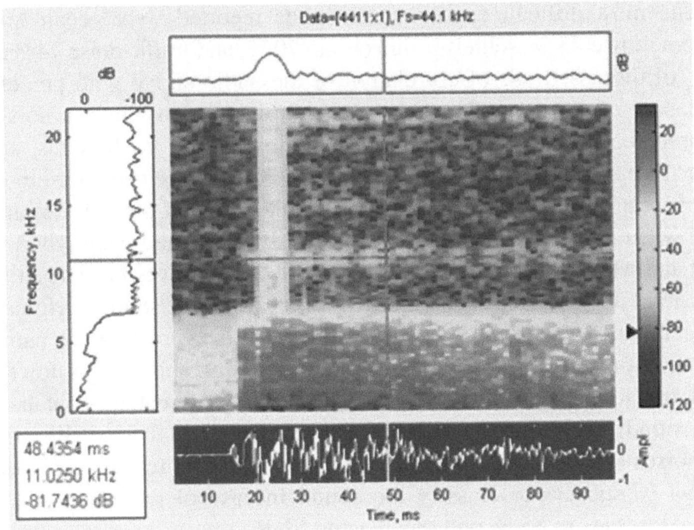

Fig. 15.1 Spectrogram of gunshot by pistol

In Fig. 15.1, it can be observed that the frequency of sound generated by pistol instantaneously increases then constant for short period (5 ms here) and after that it fall to constant amplitude.

In Fig. 15.2, it can be observed that the frequency of gunshot rises suddenly and decreases slowly. The time duration for gunshot is approximately 1 s in this case.

In Fig. 15.3, it can be observed that the frequency of gunshot by sniper decreases slowly after a peak but takes more time to decrease in comparison to the gunshot by rifle.

In Fig. 15.4, it can be observed that the frequency and intensity of gunshot remains constant for certain milliseconds, after that it starts decreasing.

These observations would help us to identify the event (gunshot) by looking for the signature of event. The identified signature and it values are presented in Table 15.2.

15.4 Features Extraction of Sound

Generally, acoustic signals are analyzed in terms of a set of features or parameters of interest and based on the parameters, the acoustic signal is classified. In this paper, mel-frequency cepstrum coefficients (MFCC) are used to extract 22 parameters of interest to gunshot, which are listed in Table 15.1 [5, 10]. The process steps of MFCC to extract the features of sound signals are as follows:

1. Estimation of amplitude and intensity of sound.
2. Grouping of frequency into bands of equal bandwidth.

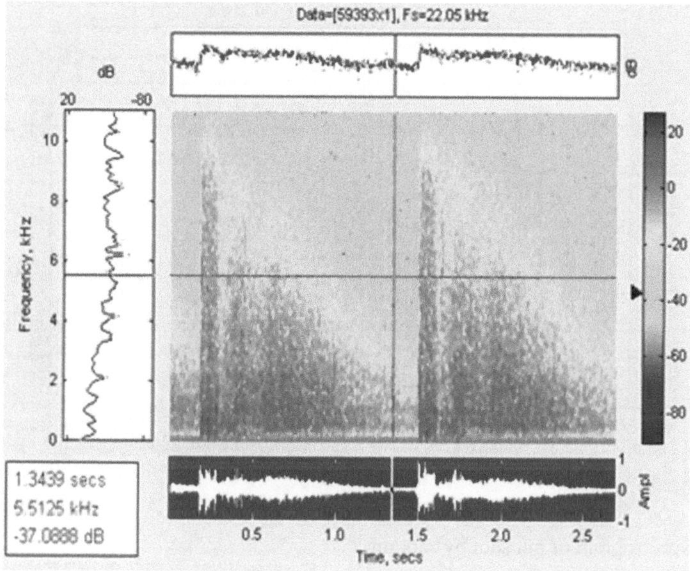

Fig. 15.2 Spectrogram of gunshot by rifle

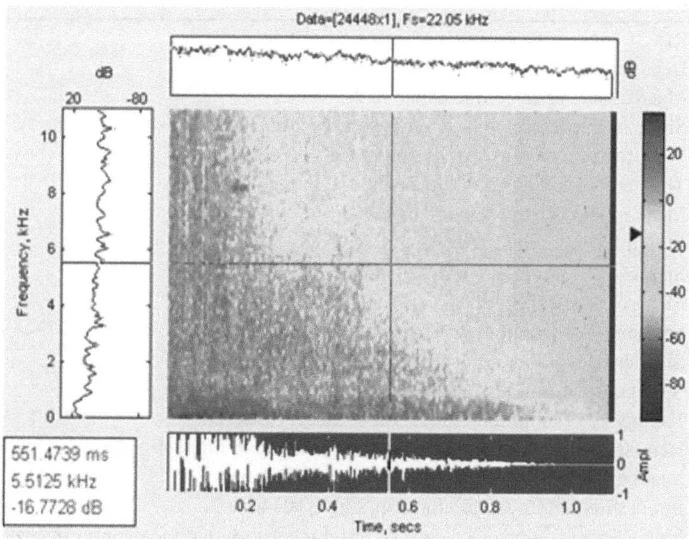

Fig. 15.3 Spectrogram of gunshot by sniper

3. Fast Fourier Transform (FFT) of bands.
4. Computation of the logarithm of bands after FFT.
5. Discrete cosine transforms (DCT) Computation of bands after computation of logarithm.

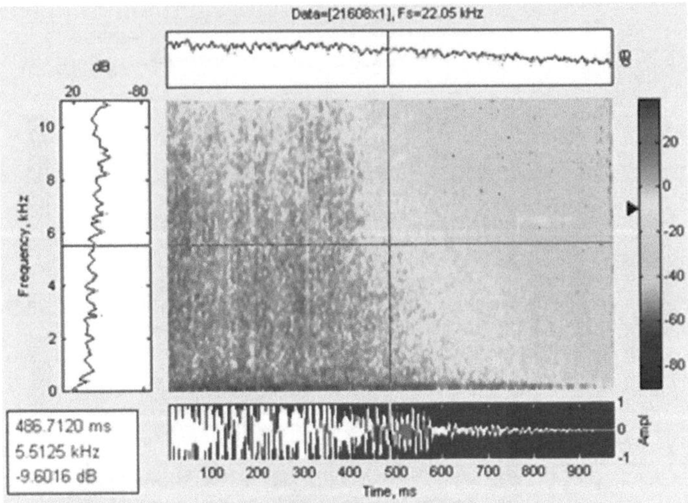

Fig. 15.4 Spectrogram of gunshot by shotgun

Table 15.1 Sound features extracted by MFCC technique

S. No.	Features description
1	Rise time, i.e., the duration of rise
2	Decay time
3	Mean square error of line adjust in 2
4	Slope of line fixed into rms-energy curve after rise
5	Crest factor, i.e., max/rms of amplitude
6	Time between the end of attack and the maximum of rms-energy
7	Mean of normalized spectral centroid
8	Maximum of normalized spectral centroid
9	Standard deviation of spectral centroid
10	Mean of spectral centroid
11	Frequency of amplitude modulation, range 4–8 Hz
12	Standard deviation of normalized spectral centroid
13	Strength of amplitude modulation, range 4–8 Hz
14	Frequency of amplitude modulation, range 10–40 Hz
15	Heuristic strength of the amplitude modulation in range 4–8 Hz
16	Standard deviation of rise times at each Bark band
17	Strength of amplitude modulation, range 10–40 Hz
18	Mean error of fit between each of onset intensities and mean onset intensity
19	Mean error of the fit between each of steady-state intensities and mean steady-state intensity
20	Overall variation of intensities at each band
21	Average cepstral coefficients during onset
22	Standard deviation of fundamental frequency

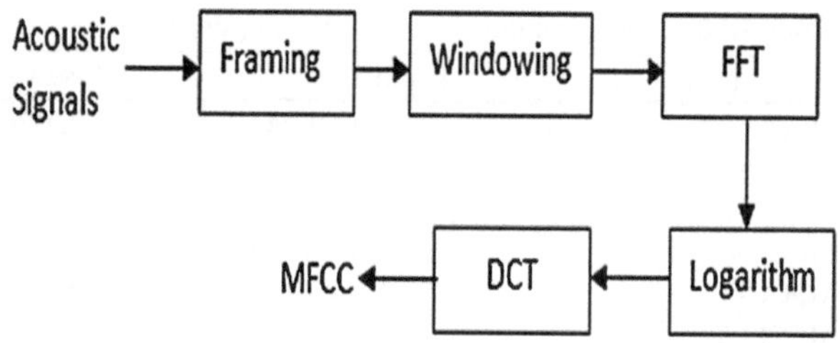

Fig. 15.5 Block diagram of MFCC process [5]

The process of parameter extraction using MFCC is shown in Fig. 15.5 as block diagram.

15.5 Training of System

The values of 22 parameters, that were discussed in the previous section were extracted from each sample signal using MFCC and are used to create a training set. Table 15.2 shows those 22 parameters that were extracted from three samples—two for pistols and one for rifle gunshot.

The extracted parameters were used for training gunshot recognition system using three-layer artificial neural networks (ANN).

The ANN uses back propagation model with momentum (Pattern Mode) learning rule. The input consists of 22 nodes, each node corresponding to one MFCC parameters, three hidden layers and one output node. The hidden layer has 11, 7, and 2 neurons Fig. 15.6. The Normalization of MFCC parameters was done in the range of 0.1–0.9.

15.6 Association Rules Learning

Association rules learning [11, 12] is a method to discover relationship between parameters of a large database. As an example, three parameters like tip in a hotel, quality of food, and quality of service could be linked through a relationship using association rules such as:

Good food + good service = good tip,
Good food + bad service = average tip,
Bad food + bad service = no tip.

After training of parameters, association rules make simple rules for decision making. Based on the association rules a decision tree (Fig. 15.7) is created for gunshot.

According to decision tree, decision rules were decided for gunshot detection.

The selected rules are the following:

Table 15.2 MFCC parameters extraction of gunshots

MFCC parameters	Pistol 1	Pistol 2	Rifle
MFCC 1	0.081929043	0.076333389	0.103508487
MFCC-2	0.345215906	0.317205015	0.450695428
MFCC-3	−0.024371645	−0.020849554	−0.016643999
MFCC-4	−0.005237523	−0.005423513	−0.004682361
MFCC-5	0.017424131	0.004221518	0.007569891
MFCC-6	0.01241868	0.008793181	0.001160881
MFCC-7	0.518305829	0.490094856	0.611755731
MFCC-8	0.032149527	−0.00358984	0.003835878
MFCC-9	−0.028921397	−0.021176352	−0.012461979
MFCC-10	−0.002598315	0.004691264	0.000558011
MFCC-11	−0.006948696	−0.001085565	−0.009483835
MFCC-12	0.521672874	0.494138025	0.615330665
MFCC-13	0.034800541	−0.002446353	0.003520145
MFCC-14	−0.028187497	−0.018725576	−0.011411904
MFCC-15	−0.001422691	0.004074864	0.001267848
MFCC-16	−0.00527637	−0.002008294	−0.01077973
MFCC-17	0.524168656	0.49807753	0.617844461
MFCC-18	0.03571343	−0.001066534	0.001942664
MFCC-19	−0.026175635	−0.014829641	−0.010374597
MFCC-20	0.000199536	0.004610543	0.00075864
MFCC-21	−0.002661296	0.000608799	−0.00731112
MFCC-22	0.040353661	0.040353661	0.039908543

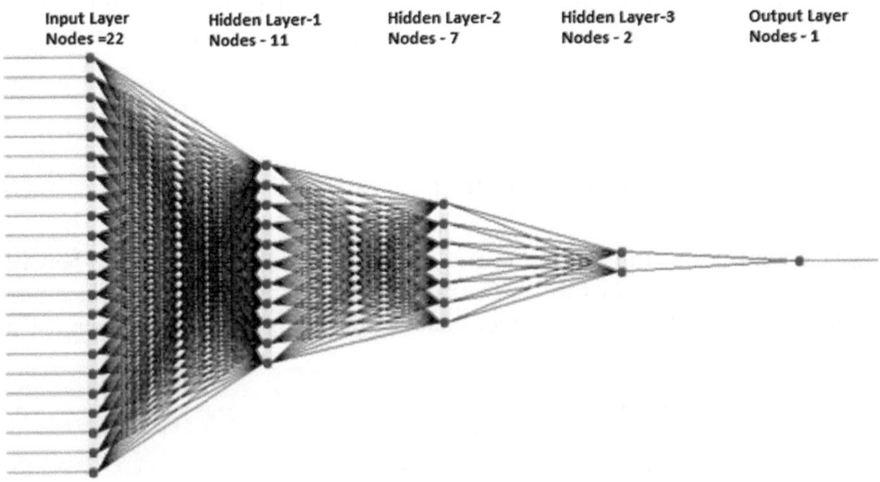

Fig. 15.6 Architecture of training model (Neurons: 22-11-7-2-1)

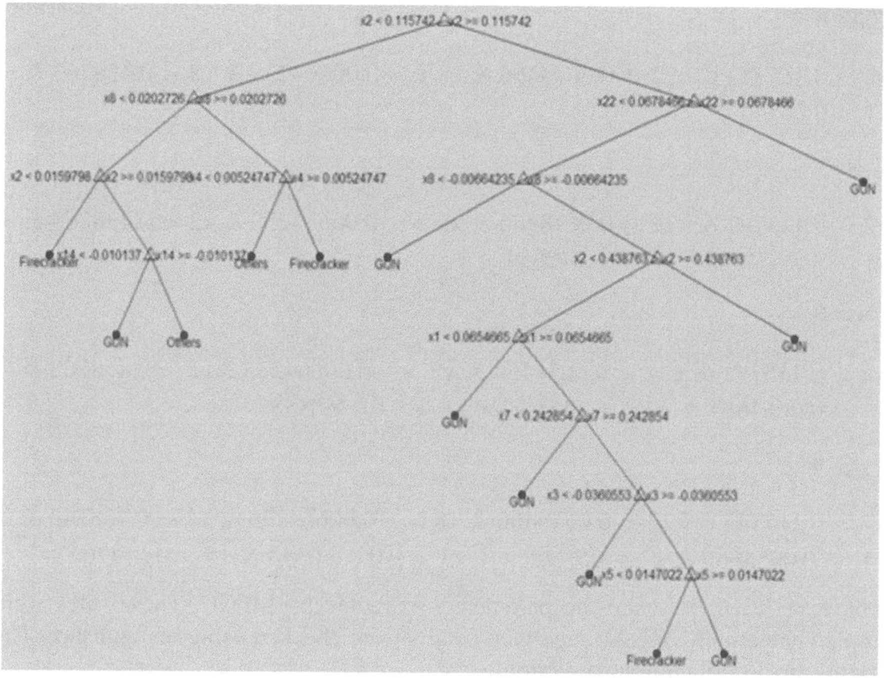

Fig. 15.7 Decision tree for gunshot detection based on MFCC parameters

Case 1

$$x2 < 0.115742 \ \& \ x8 < 0.0202726 \ \& \ x2 \geq 0.01597984 \ \& \ x14 < -0.010137$$

Case 2

$$x2 > 0.115742 \ \& \ x22 > 0.0678466$$

Case 3

$$x2 > 0.115742 \ \& \ x22 < 0.0678466 \ \& \ x8 < -0.00664235$$

Case 4

$$x2 > 0.115742 \ \& \ x22 < 0.0678466 \ \& \ x8 > -0.00664235 \ \& \ x2 > 0.438763$$

Case 5

$$x2 > 0.115742 \ \& \ x22 < 0.0678466 \ \& \ x8 > -0.00664235 \ \& \ x2 > 0.438763$$

Case 6

$x2 > 0.115742$ & $x22 < 0.0678466$ & $x8 > -0.00664235$ & $x2 < 0.438763$ & $x1 < 0.0654665$

Case 7

$x2 > 0.115742$ & $x22 < 0.0678466$ & $x8 > -0.00664235$ & $x2 < 0.438763$ & $x1 > 0.0654665$ & $x7 < 0.242854$

Case 8

$x2 > 0.115742$ & $x22 < 0.0678466$ & $x8 > -0.00664235$ &$x2 < 0.438763$ & $x1 > 0.0654665$ & $x7 > 0.242854$ & $x3 < -0.0360553$

Case 9

$x2 > 0.115742$ & $x22 < 0.0678466$ & $x8 > -0.00664235$ & $x2 < 0.438763$ & $x1 > 0.0654665$ & $x7 > 0.242854$ & $x3 > -0.0360553$ & $x5 > 0.0147022$

where $x1$, $x2$,…, $x22$ are the parameters extracted by MFCC. These rules were implemented on MATLAB, and then results were checked using real gunshot. The results are described in next section.

15.7 Testing in the Forest Area

The gunshot recognition algorithm was tested in a forest area. A Microphone connected to a laptop was used to capture the acoustic signals. Forest clutter and considerable noise as clapping was present when gunshot was fired and was detected successfully. The Graphical user interface (GUI) of the algorithm is shown in Fig. 15.8. The start button is to enable the system for gunshot detection. It can be put in continuous monitoring mode also, in which it will record and process the recorded sound simultaneously. The spectrogram of the acoustic signal is plotted as a graph. The area below the graph shows the result of detection. It displays either "Gunshot Detected" or "No Gunshot."

Figure 15.9 shows the gunshot detection in heavy noise. The axis in GUI shows spectrogram of both gunshot and noise together. And it displays result as "Gunshot Detected."

The algorithm is efficient to detect gunshot in forest environment; the actual gunshot testing is 95 % efficient with environmental noise which decreases up to 85 % decreased SNR. It generates 5 % false positive result for firecrackers (bomb). Algorithm has been tested with 70 gunshots in which 66 gunshots were detected. 50 gunshots were tested in the presence of noise in which 42 gunshots were detected.

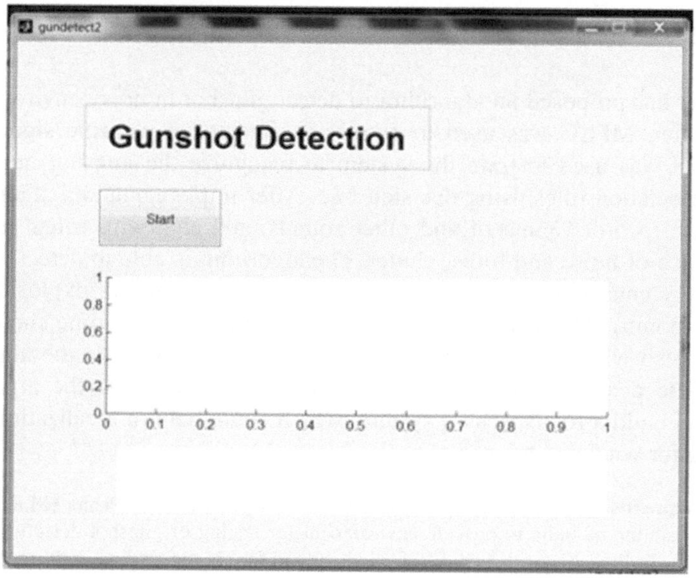

Fig. 15.8 GUI of gunshot detection system

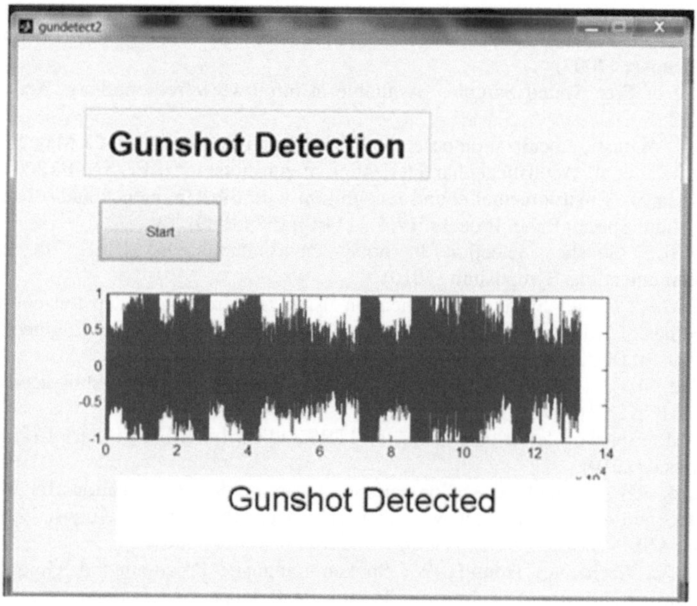

Fig. 15.9 GUI of gunshot detection in heavy surrounding noise

15.8 Conclusion and Future Work

This paper has proposed an algorithm to detect gunshot in noisy environment and forest clutter. MFCC was used to extract parameters of acoustic signals, three-layer ANN was used to train the system to recognize the gunshot implemented by the association rules using decision tree. After implementation, algorithm was tested with recorded gunshot and other sounds, and also with actual gunshot in the presence of noise and forest clutter. The algorithm is able to detect gunshot in heavy noisy environment. It generates false positive results for explosion of fire-crackers (bomb). The system gives false negative results for some shots of shot-gun, and also when SNR is low. The future work has to refine the algorithm to reduce false positives and remove false negatives, implement the algorithm on FPGA and build a robust sensor system, which could help in localization by compensating for wind and temperature effects.

Acknowledgments We thank the officials of Panna Tiger Reserve, Panna (M.P., India) and Wild Life Institute of India to provide environment for testing of gunshot detection algorithm. We also thank to Prof. Mehar Kayal, EPFL, Switzerland for his suggestions and support.

References

1. Graupe, D.: Principles of Artificial Neural Networks. World Scientific Publishing Co. Pte. Ltd, Singapore (2007)
2. Database of Free Sound Samples. Available at http://www.freesound.org. Accessed on 26 July 2012
3. Aleksi, I.: Acoustic Localization based on FPGA. Opatija, Croatia, 24–28 May 2010
4. Maher, R.C., et al.: Acoustical characterization of gunshots. In: IEEE SAFE (2007)
5. Chu, S., et al.: Environmental sound recognition with time frequency audio features. IEEE Trans. Audio Speech Lang. Process. **17**(6), 1142–1158 (2009)
6. Freire, IL.: Gunshot detection in noisy environments. In: IEEE 7th International Telecommunications Symposium (2010)
7. Hasan, M.R., et. al.: Speaker identification using mel frequency Cepstral coefficients. In: Proceedings of 3rd International Conference on Electrical and Computer Engineering (ICECE 2004), 28–30 Dec 2004
8. Ghiurcau, M.V., et al.: Wildlife intruder detection using sounds captured by acoustic sensors. In: IEEE ICASSP (1992)
9. Smith, M.: Gunshot detection system for JTRS radios. In: IEEE Military Communications Conference (2010)
10. Davisand, S.B., Mermelste, P.: Comparison of parametric representations for monosyllabic word recognition in continuously spoken sentences. IEEE Trans. Acoust. Speech Signal Process. **ASSP-28**(4), 357–366 (1980)
11. Huang, X., Acero, A., Hon, H.-W.: Spoken Language Processing: A Guide to Theory, Algorithms, and System Development. Prentice Hall, Upper Saddle River (2001)
12. Tan, P.N., Steinbach, M., Kumar, V.: Chapter 6. Association Analysis: Basic Concepts and Algorithms: Introduction to Data Mining. Addison-Wesley, Boston (2005). ISBN 0-321-32136-7

Chapter 16
Fault Tolerant QoS Adaptive Clustering for Wireless Sensor Networks

T. Shiva Prakash, K. B. Raja, K. R. Venugopal, S. S. Iyengar and L. M. Patnaik

Abstract This paper proposes and analyzes an Energy Efficient Fault Tolerant QoS Adaptive Clustering Algorithm (FTQAC) for Wireless sensor networks suitable to support real-time traffic. The protocol achieves fault tolerance and energy efficiency through a dual cluster head mechanism and guarantees the desired QoS by including delay and bandwidth parameters in the route selection process. Simulation results indicate that FTQAC reduces overall energy consumption and improves network lifetime while maintaining required QoS.

Keywords Clustering • Energy efficiency • Fault tolerance • Packet delivery ratio (PDR) • Quality of service (QoS) • Wireless sensor networks

16.1 Introduction

Wireless sensor networks (WSNs) can be defined as the representative noninfrastructure networks that are capable of wireless communication. Emerging WSNs have stringent QoS requirements that include fault tolerance, timeliness, and reliability. The timeliness and reliability level for data exchanged between sensors and base station is of paramount importance especially in real-time scenarios. Thus, QoS routing is an important topic in sensor networks research, hence there is a

T. Shiva Prakash (✉) · K. B. Raja · K. R. Venugopal
University Visvesvaraya College of Engineering, Bangalore, India
e-mail: shivprakasht@gmail.com

S. S. Iyengar
Florida International University, Miami, FL, USA

L. M. Patnaik
Indian Institute of Science, Bangalore, India

R. Maringanti et al. (eds.), *Proceedings of Ninth International Conference on Wireless Communication and Sensor Networks*, Lecture Notes in Electrical Engineering 299, DOI: 10.1007/978-81-322-1823-4_16,

growing interest in the literature on proposals for QoS routing in WSNs. In order to satisfy the QoS requirements and energy constraints for WSNs, hierarchical (clustering) techniques have been an attractive approach to organize sensor networks based on their power levels and proximity.

Motivation: The cluster-based network model provides inherent optimization capabilities at cluster heads such as data fusion and reduce communication interference by using time division multiple access (TDMA). High-energy nodes can be used to process and send the information while low-energy nodes can be used to perform the sensing task. While earlier works (explained in the Sect. 16.2) were primarily focused on the above-mentioned aspects, more recent research have begun to consider fault tolerance, reliability, and quality-of-service and our proposed protocol is motivated by these metrics.

Contribution: This proposed protocol Fault Tolerant QoS Adaptive Clustering Algorithm (FTQAC) employs an adaptive fault tolerant dual cluster head mechanism in the cluster with respect to the working of the cluster head and guarantees the desired QoS by including delay and bandwidth parameters in the route selection process. Furthermore, the protocol evenly distributes the energy consumption to all nodes so as to extend the sensor network lifetime. We test the performance of our proposed approaches by implementing our algorithms using *ns*-2 simulator. Our results demonstrate the performance and benefits of our algorithm.

The rest of the paper is organized as follows: Sect. 16.2 gives a review of Related Works. Sections 16.3 and 16.4 explain the Network Model, notations, and assumptions and the algorithm. Section 16.5 presents the Simulation and Evaluation of the algorithm. Conclusions are presented in Sect. 16.6.

16.2 Related Work

In this section, a summary of the current state of the art in hierarchical routing protocols for WSNs are presented. Low-Energy Adaptive Clustering Hierarchy (LEACH) [1] is a self-organizing, adaptive clustering-based protocol that uses randomized rotation of cluster heads to uniformly distribute the energy load among the sensor nodes in the network. Threshold sensitive energy efficient sensor network (TEEN) and its adaptive version (AdaPtive) threshold sensitive energy efficient sensor network (APTEEN) [2] are clustering protocols that are comparable to LEACH, they are sensitive to sudden changes in WSNs. Two-Level Hierarchy LEACH (TL-LEACH) [3] algorithm elects two sensor nodes in each cluster as cluster heads, one node as primary cluster head and the other as the secondary cluster head.

Chen et al. [4], developed a local centralized method for electing a dual cluster head and introduce a parameter to measure the QoS support in hierarchical applications of WSNs. The dual cluster head model can also improve the survivability of wireless sensor networks. The shortcoming of the protocol is that the secondary cluster is created only if the number of nodes in a given cluster is greater than a threshold, the protocol proposed in this paper always creates a secondary cluster to achieve fault tolerance in the WSN. Muruganathan et al. [5], proposed a base station-controlled dynamic clustering protocol (BCDCP), which utilizes the high-energy base station to perform most energy intensive tasks and distributes the energy dissipation evenly among all sensor nodes to improve network lifetime. Peng et al. [6], propose a protocol that focuses on improving the energy efficiency and other QoS parameters by excluding the node with improper geographic location to be the cluster heads.

Hossein et al. [7], use cluster heads as higher power relay nodes in a two-tiered WSN and these relay nodes may form a network among themselves to route data toward the sink and provide energy efficient QoS routing in cluster-based WSNs. Aslam et al. [8], use Network Calculus to present a mathematical model of a TDMA-based medium access control protocol, where a cluster-based system is modeled and arrival/service curve is proposed. Shiva Prakash et al. [9] paper proposes a Traffic-Differentiated Two-Hop Routing protocol the protocol achieves to increase packet reception ratio (PRR) and reduce end-to-end delay while considering multi-queue priority policy, two-hop neighborhood information, link reliability, and power efficiency. Fapojuwo et al. [10], designed a Quality of service enhanced base station controlled dynamic clustering protocol (QBCDCP). The protocol achieves energy efficiency through a rotating head clustering mechanism and delegation of energy intensive tasks to the base station, while providing QoS support by including delay and bandwidth parameters in the route selection process. In the proposed protocol, we employ a dual cluster head model to attain fault tolerance and improve the lifetime of the WSN, also the dual cluster head model aids in enhancing the end-to-end delay and packet delivery ratio (PDR).

16.3 Problem Definition

The topology of a wireless sensor network may be described by a graph $G = (N, L)$, where N is the set of nodes and L is the set of links. The objectives are to,

- Improve the lifetime of the network.
- Reduce the average end-to-end packet delay.
- Minimize the packet delivery ratio (PDR).

16.3.1 System Model and Assumptions

In our system model, we assume the following:

- The wireless sensor network consists of N homogeneous sensor nodes, deployed at random locations in a sensor field. An example scenario is shown in Fig. 16.1, where the sensor field is a square area at a distance d_{BS} from a single fixed base station. The sensors are grouped into 1-hop clusters with a specific clustering algorithm. All sensor nodes are immobile and are powered by a constant nonrenewable on-board energy source.

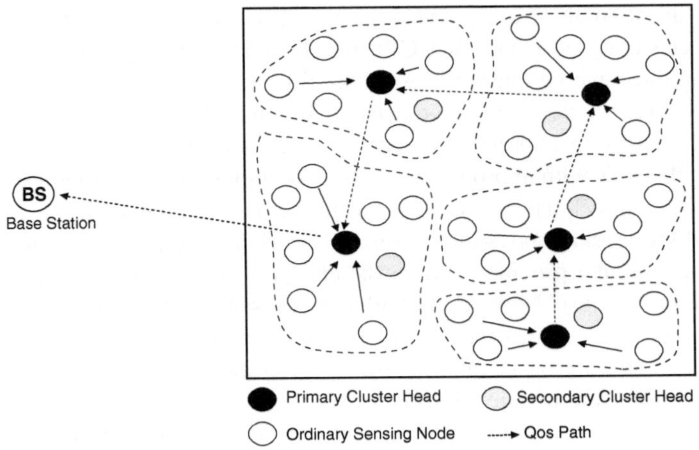

Fig. 16.1 System model

- All nodes are supposed to be aware of their residual energy and are capable of measuring the signal strength indicator (RSSI) of a received message, this measurement may be used as an indication of distance from the sender.
- The nodes in a cluster may perform either of three roles: primary cluster head, secondary cluster head, or sensing. Each cluster head performs activities such as scheduling of intra-cluster and inter-cluster communications, data aggregation, and data forwarding to the base station through multi-hop routing. The role of the secondary cluster head is to emulate the role of the primary cluster head in case of its failure. On the other hand, a sensing node may be actively sensing the target area.
- The cluster head, gathers data from the other nodes within its cluster, performs data aggregation/fusion, and routes the data to the base station through other cluster head nodes. The base station in turn performs the key tasks of cluster formation, cluster head selection, and cluster head to cluster head QoS routing path construction.
- The base station has knowledge via internal global positioning system (GPS) of the position of all nodes inside the sensor field. The base station has a constant power supply and thus, has no energy constraints.
- Radio Model: The energy required at the transmitter amplifier to guarantee an acceptable signal level at the receiver, when receiver and transmitter are separated by a distance *d*, $E_a(d)$ is:

$$E_a(d) = \begin{cases} \varepsilon_{FS} d^2, d \leq d_o \\ \varepsilon_{TR} d^4, d \leq d_o \end{cases}, \quad d_o = \sqrt{\frac{\varepsilon_{FS}}{\varepsilon_{TR}}} \tag{16.1}$$

where $\varepsilon_{\mathrm{FS}} d^2$ and $\varepsilon_{\mathrm{TR}} d^4$ denote the transmit amplification parameters corresponding to the free-space and two-ray models, respectively, and d_o is the threshold distance.

16.4 Algorithm

The proposed protocol FTQAC incorporates QoS requirements like fault tolerance, delay, and bandwidth information during route establishment. The operation of the protocol is split into phases. The first stage of FTQAC consists of the cluster splitting and primary cluster selection the second phase involves the selection of secondary cluster head. The last phase involves formation of the QoS route from cluster head to the base station. Time division multi-access (TDMA) and spreading code are engaged to minimize inter-cluster interference to allow simultaneous transmissions in neighboring clusters.

16.4.1 Cluster Setup and Primary Cluster Head Selection

In the proposed protocol, the cluster splitting and primary cluster head selection is accomplished by the Base Station as in [5].

16.4.2 Secondary Cluster Head Selection

In the next phase, the primary cluster head (PCH) has the role of identifying the secondary cluster head (SCH), the steps involved are shown below.

1. Each new PCH sends message M_1 to the sensing nodes in the cluster, the message contains the node's ID and a header to distinguish the message.
2. The sensing nodes record the Received Signal Strength Indicator (RSSI) of message M_1. The sensing nodes send message M_2 to the PCH. The message contains the node's ID, ID code of the PCH, RSSI value of message received from the PCH, and the current residual energy of the node.
3. The PCH receives M_2 from ordinary nodes, the cluster head calculates the average residual energy level of all sensing nodes in the cluster. It selects a SCH from one of the nodes which has the largest RSSI of message M_1 among the qualified nodes whose residual energy is more than the average residual energy of all nodes in the cluster.
4. The PCH sets up TDMA schedule and transmits the schedule to the SCH and the sensing nodes in the cluster. The role of the SCH is to emulate the PCH in case of its failure. The PCH sends a message M_3 periodically to the SCH informing its role and its current residual energy status. The SCH sends a ACK back to the PCH.
5. When the residual energy of the PCH is equal to or less than the E_t (threshold energy level) the PCH relinquishes its role to the SCH by sending a common message to all nodes in cluster. The new primary cluster updates the base

station of its delay, bandwidth and residual energy of the sensing nodes, it continues the functions of the cluster head using the same TDMA schedule.

6. The base station triggers reclustering process only when more than one-third of the SCHs have reached their E_t, also it assigns the new E_t level for the next round based on the average residual energy of selected PCHs. This process prevents frequent reclustering and avoids excessive depletion of the cluster heads battery, this mechanism results in better power efficiency.

16.4.3 QoS Route Establishment

Algorithm 1: QoS Route Establishment

Phase III of Fault Tolerant QoS Adaptive Clustering Algorithm (FTQAC)
Input: C (Set of primary cluster heads (PCHID)),
DPCHID (Destination Primary Cluster Head ID), BW_{req} (Minimum bandwith required), D_{req} (End-to-end delay required), BW_{xy} (Bandwith offered by link xy), D_{xy} (Delay associated with link xy), $E_a(d_{xy})$ (Power Amplifier energy of current cluster head, which is a function of the distance between the cluster heads and radio propagation model)
Output: Optimal QoS Path from Base Station to requesting Primary Cluster Head

```
for each PCHID ∈ C do
    if BW_xy ≥ BW_req then
        D_Sum = D_Sum + D_xy ;
        E_aSum = E_aSum + E_a(d_xy) ;
        Add PCHID to R;
        if PCHID == DPCHID then
            if D_Sum ≤ D_req then
                Add the path R and E_aSum of path to Q_R;
            else
                Discard R;
        else
            continue;
    else
        Discard R;
for each R ∈ Q_R do
    Return Q_S with Min {E_aSum}
return Q_S;
```

The desired QoS metrics for route establishment, i.e., delay, bandwidth of cluster head nodes, and residual energy of the sensing nodes are aggregated and reported to the base station periodically. Delay and bandwidth are measured at cluster head nodes. The delay associated to traversing a particular cluster head is, the time duration between entering the input queue and leaving the output queue of the cluster head (D_{xy}). Bandwidth is computed at each cluster head as the number of free time slots within each cluster head (BW_{xy}). When a connection is desired, the base station sets up a QoS-based route Q_S between the cluster head where the connection is initiated and itself as shown in Fig. 16.1. The base station finds the route which minimizes the delays and power amplifier energy along the path, and has a minimum bandwidth greater than or equal to the requested bandwidth (BW_{req}) as shown in Algorithm 1. The algorithm may produce more than one optimal path, the path having cluster heads with minimum power amplifier energy (E_{aSum}) is chosen. After a route is chosen, the base station communicates it to the concerned cluster head nodes, which schedule the connection by specifying the required number of time slots to maintain it. During the communication phase when the PCH is depleted of energy it transfers its role to the SCH. But, the PCH is currently involved in the QoS path hence it informs both the downstream cluster head, upstream cluster head, and the base station of its duty transfer and then relinquishes its role. The traffic is redirected to the new PCH and the QoS level is maintained throughout the duration of the connection.

16.5 Performance Evaluation

To evaluate the proposed protocol, we carried out a simulation study using ns-2 a discrete event simulator. The proposed protocol FTQAC is compared with QBCDCP. The simulation configuration consists of 100 nodes where each node is assigned an initial energy of 2 J, located in a 100 m^2 area. The base station is located 25 m from the sensor field. The end-to-end delay objective D_{req} is fixed at 10 s and BW_{req} was set at 16 Kbps by assigning each connection one out of 16 available TDMA time slots. A comparison of the average residual energy of cluster heads and packet delivery ratio (PDR) for different loads are obtained for FTQAC and QBCDCP.

In QBCDCP during the communication phase if the primary cluster head is depleted of energy, the entire cluster does not function and causes the WSN to become unstable and inconsistent. This problem can be overcome by the dual cluster head model. In FTQAC the cluster will continue to work reliability since the SCH takes the role of the PCH when the threshold (E_t) energy is reached. In QBCDCP the cluster formation is triggered frequently since the cluster head gets depleted of energy quickly. In Fig. 16.2a, the characteristics of both the protocols are similar initially since the energy level of the cluster heads are high, but during he later stage of simulation the average residual energy of PCH in FTQAC is higher since the PCH relinquishes its role to the SCH. This model of dual cluster head has the feature of adaptive fault-tolerance and improves the robustness of the WSN. From Fig. 16.2a it is observed that there is about 15 % increase in the network lifetime using the dual cluster head model.

As depicted in Fig. 16.2b the packet delivery ratio (PDR), decreases as the packet arrival rate increases. It is observed that FTQAC performs marginally better than QBCDCP when the packet arrival rate is above 30 packets per second. In QBCDCP as the packet arrival rate increases the cluster head in the QoS path will get depleted of energy and the connection will be terminated, triggering route repair, and hence result in a lower PDR. In FTQAC, the role transfer from PCH to SCH will ensure that the scheduled connection will not be dropped hence maintaining the packet delivery ratio.

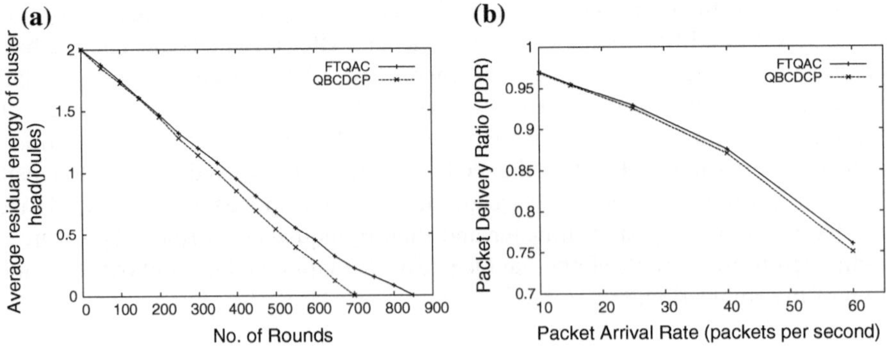

Fig. 16.2 Performance evaluation: FTQAC versus QBCDCP, (**a**) no. of rounds versus average residual energy of cluster, (**b**) packet arrival rate versus packet delivery ratio

16.6 Conclusions

The FTQAC protocol achieves fault tolerance through a dual cluster head mechanism and guarantees the desired QoS using bandwidth, delay, and transmission energy metrics. The FTQAC provides an improvement of up to 15 % in lifetime over QBCDCP, while obtaining similar end-to-end delay objective over different loads. The FTQAC is a feasible solution to the QoS routing problem in power constrained wireless sensor networks.

References

1. Heinzelman, W.R., Chandrakasan, A., Balakrishnan, H.: Energy efficient communication protocol for wireless microsensor networks. In: Proceedings of the 33rd Annual Hawaii International Conference on System Sciences, vol. 2. pp. 2–11 (2000)
2. Manjeshwar, A., Agrawal, D.P.: A hybrid protocol for efficient routing and comprehensive information retrieval in wireless sensor networks. In: Proceedings of 2nd International Workshop on Parallel and Distributed Computing Issues in Wireless Networks and Mobile Computing, pp. 195–202. Ft. Lauderdale (2002)
3. Loscri, V., Marano, S., Morabito, G.: A two-levels hierarchy for low-energy adaptative clustering hierarchy (TL-LEACH). In: Proceedings of VTC2005, pp. 1809–1813. Dallas, USA (2005)
4. Chen, W., Li, W., Shou, H., Yuan, B.: A QoS-based adaptive clustering algorithm for wireless sensor networks. In: Proceedings of the 2006 IEEE International Conference on Mechatronics and Automation, pp. 1947–1952. Luoyang, China (2006)
5. Muruganathan, S.D., Ma, D.C.F., Bhasin, R.I., Fapojuwo, A.O.: A centralized energy-efficient routing protocol for wireless sensor networks. IEEE Commun. Mag. **43**(3), S8–S13 (2005)
6. Ji, P., Wu, C., Zhang, Y., Chen, F.: A low-energy adaptive clustering routing protocol of wireless sensor networks. In: Proceedings of International Conference on Wireless Communications, Networking and Mobile Computing (WiCOM), pp. 1–4 (2011)

7. EkbataniFard, G.H., Monsefi, R., Akbarzadeh-T, M.R., Yaghmaee, M.H.: A multi-objective genetic algorithm based approach for energy efficient QoS-routing in two-tiered wireless sensor networks. In: Proceedings of International Symposium on Wireless Pervasive Computing (ISWPC), pp. 80–85 (2010)
8. Aslam, N., Phillips, W., Safdar, G.A.: Worst case bounds of a cluster-based MAC protocol for wireless sensor networks. In: Proceedings of Wireless Telecommunications Symposium (WTS), pp. 1–6 (2012)
9. Shiva Prakash, T., Raja, K.B., Venugopal, K.R., Iyengar, S.S., Patnaik, L.M.: Traffic-differentiated two-hop routing for QoS in wireless sensor networks. In: IEEE Proceedings of the 5th International Conference on Cyber-Enabled Distributed Computing and Knowledge Discovery (CyberC), Beijing, China, 10–13 Oct 2013
10. Fapojuwo, A.O., Cano-Tinoco, A.: Energy consumption and message delay analysis of QoS enhanced base station controlled dynamic clustering protocol for wireless sensor networks. IEEE Trans. Wireless Commun. **8**(10), 5366–5374 (2009)

Chapter 17
A New Approach for Data Filtering in Wireless Sensor Networks

Nidhi Gautam, Sanjeev Sofat and Renu Vig

Abstract Wireless Sensor Network has encouraged researchers to broaden up the field by critically evaluating and realizing its capabilities in various application areas. With every innovation there comes along lot of challenges. Conceiving an idea of implementing wireless sensor network has shown many challenges, i.e., node deployment, data clustering, data aggregation, energy efficiency, lifetime improvement, etc. In this paper, we have proposed a filtering scheme at the sensor/relay node which filters out the spurious and redundant data and is applicable both for critical as well as noncritical applications. The proposed approach shows promising results by filtering out useless data. This technique improves energy efficiency and network lifetime of the network.

Keywords Filtering • Clustering • Aggregation • Network lifetime • Energy consumed • Delay

17.1 Introduction

Wireless sensor network (WSN) is intended to be deployed in tough and harsh environments where sensors are exposed to herculean conditions. Energy efficiency and lifetime are the major concerns in WSN due to resource constraint battery life.

N. Gautam (✉)
U. I. A. M. S. Panjab University, Chandigarh, India
e-mail: nidhig121@gmail.com

S. Sofat
PEC University of Technology, Chandigarh, India
e-mail: sanjeevsofat@pec.ac.in

R. Vig
U. I. E. T. Panjab University, Chandigarh, India
e-mail: renuvig@hotmail.com

R. Maringanti et al. (eds.), *Proceedings of Ninth International Conference on Wireless Communication and Sensor Networks*, Lecture Notes in Electrical Engineering 299, DOI: 10.1007/978-81-322-1823-4_17,

In-network processing, aggregation, data fusion, clustering has helped to overcome such issues to some extent. Despite so many alternatives, still there exists scope for improvement. Data filtering and fusion is the key approach in in-network data processing which is used to combine data from multiple sources in such a way so that data does not lose its granularity. This technique fuses the data from all the sources and hence reduces the size. It is suitable both for centralized as well as for distributed systems. In centralized system, raw data is sent by sensor nodes and data fusion will be at the centralized node. In distributed system, various data fusion techniques can be implemented on distributed components. Hence, it is a multilevel process that deals with sensing data, estimating, and associating the correlated data from several sources. The data filtering and fusion appeared to be useful in reducing communication overhead by banishing redundant messages. Hence, it seems to be a very critical factor in increasing network lifetime and energy efficiency of WSN [1].

17.2 Literature Review

The algorithms presented by Du et al. (2006) and Zhou et al. (2006) presented algorithms LBDAT [2] and HDA [3], respectively. Both of these algorithms are based on hierarchical approach for data aggregation and do not implement any data fusion process to reduce data size. The algorithm AFST presented by Luo et al. [4] checks if fusion leads to improve energy then only fusion will be performed otherwise data will be relayed to the sink node as it is. Verdone et al. (2008) discussed about Parallel fusion architecture (PFA) and cooperative fusion architecture (CFA) in [5]. Zhu et al. [6] proposed a correlation aware aggregation approach SCT. Hua et al. [7] proposed a model to optimize routing as well as aggregation simultaneously. Luo et al. [8] suggested that chain and tree structures were more optimal with minimum energy reliability information gathering (MERIG). Xu et al. [9] reported a model on noisy WSN based on a Gaussian WSN with an objective of minimizing mean distortion level of recovered data through least square error (LSE) and proposed a cross-layer deign to optimize the fusion performance, energy consumption, and delay requirements. Renjith et al. [10] presented a survey of various data aggregation approaches for wireless sensor networks.

Zhang et al. [11] used Kalman filter at the leaf nodes as well as at the relay nodes to find out the optimal fusion and predicts the upper stream sensor's data that cannot be aggregated to the sink before deadlines. Ren et al. [12] proposed ADA, an attribute aware data aggregation scheme with a concept of packet attribute with potential field from physics and pheromone from ant colony systems. Cheng et al. [13] proposed a delay-aware network for WSNs with in-network data fusion with a tree-structured network in which sensor nodes are organized into multiple single layer clusters of different sizes so that they can communicate with fusion center in an interleaved manner. Lu et al. [14] proposed a distributed data fusion routing protocol to find optimal paths for a given distributed data fusion tree that consists of three steps: path discovery, path maintenance, and path adaptation. Lu et al. [15] in another paper proposed both optimal and approximate solutions to map application task graph to network nodes and place fusion functions to achieve energy efficiency. In this paper, the author aimed to minimize energy cost of distributed data fusion application deployed in Active Networks. Khaleghi et al. [16] surveyed various data fusion approaches in wireless sensor networks. Sharma et al. [17] proposed a TWSW filtering approach but it does not cover criticality of the application and is majorly suitable for noncritical applications.

There is no filtering approach which caters the requirements of both kinds of applications for energy efficiency and improving network lifetime. This paper proposed a filtering approach that is suitable for critical as well as for noncritical application. This filtering approach filters out the data in such a way that spurious and redundant data can be ignored for noncritical applications and crucial information is disseminated for critical applications. The results show that this approach is energy efficient for wireless sensor networks.

17.3 Assumptions and System Model

- Sensor nodes are homogeneous, can act as relays, aware of their locations, and placed in 2D space.
- The uniformly placed cluster head nodes have unlimited battery. They are aware of neighbor cluster head nodes and act as fusion centers.
- Sink node (BSN) in the network is stationary placed outside the area of observation with unlimited power.
- 100 sensor nodes, 10 cluster head nodes, network area $(100 \times 100)\text{m}^2$, power parameter MicaZ, ZigBee Application with 127 bytes packet size, IEEE 802.15.4 standard at the MAC and Physical Layer, Linear Battery model (1,200 mAh) for sensor nodes, two-ray signal propagation model.
- The value for N_{SI}, N_{DI}, and α is 20 s, 100 s, and ± 0.02 °C, respectively.

17.4 Proposed Algorithm

The proposed approach is a filtering window at the sensor/relay node that acts according to the severity of the application. N_{UValue} and N_{LValue} represent the upper and lower bounds, respectively of the filtering window. This approach is suitable both for critical as well as for noncritical applications. Following terminology has been used to represent our filtering approach:

$$N_{\text{UValue}} = N_{\text{m}} + N_{\text{std}} \tag{17.1}$$

$$N_{\text{LValue}} = N_{\text{m}} - N_{\text{std}} \tag{17.2}$$

$$A = \text{Average}(N_{\text{b}}) \tag{17.3}$$

$$B = \text{Average}(N_b) + \alpha \tag{17.4}$$

$$C = \text{Average}(N_b) - \alpha \tag{17.5}$$

where N_m is the mean value, N_{std} is the standard deviation, N_b is the buffer, A is average value of buffered node's data, N_{SI} is sensing interval, N_{DI} is data dissemination interval, T_t is total time of observation, α is the local deviation.

When data dissemination occurs, it checks whether past data is available to calculate N_{UValue} and N_{LValue}. For no availability of past data, data dissemination takes place and N_m, N_{UValue}, and N_{LValue} values are calculated and stored for further processing of data. If past data is available then check for the criticality of the application. For critical applications, spurious data, i.e., data beyond the upper and lower bounds are disseminated further, otherwise data dissemination is blocked. For noncritical applications, the spurious data is blocked by setting the window size between N_{UValue} and N_{LValue}. It also blocks redundant data by setting the local deviation α. Hence, our filtering approach reduces inter-node data transmissions in the network which further improves energy efficiency and network lifetime of the network. The flowchart explains the algorithm in detail for both the critical and noncritical applications.

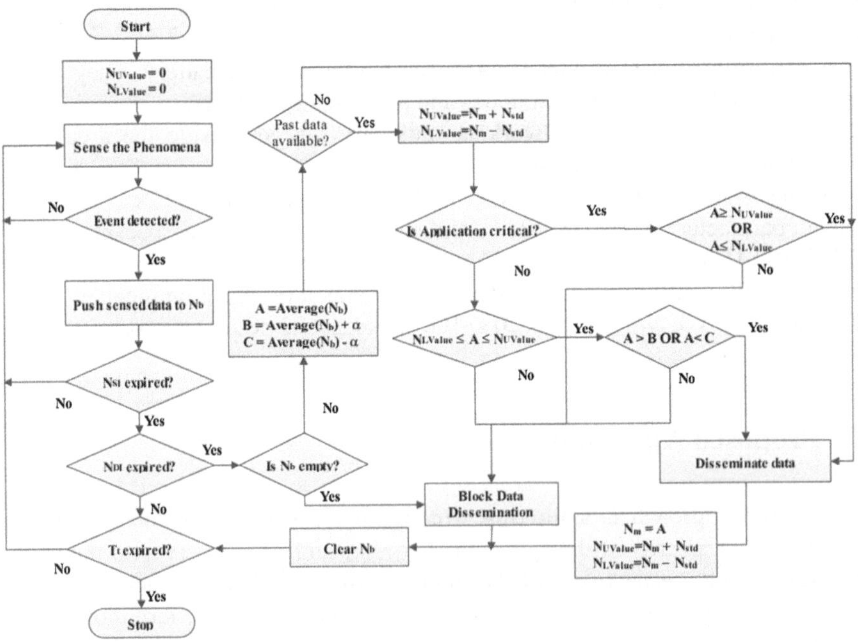

17.5 Results and Discussions

A voronoi ant systems (VAS) algorithm has been used in this work. VAS used voronoi scheme for clustering and ant systems for routing the data. VAS has proved to be better than other algorithms for wireless sensor networks [18].

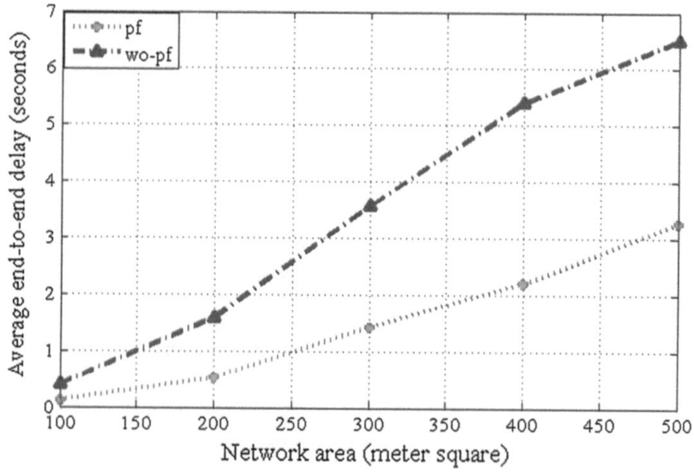

Fig. 17.1 Average end-to-end delay with varying network area

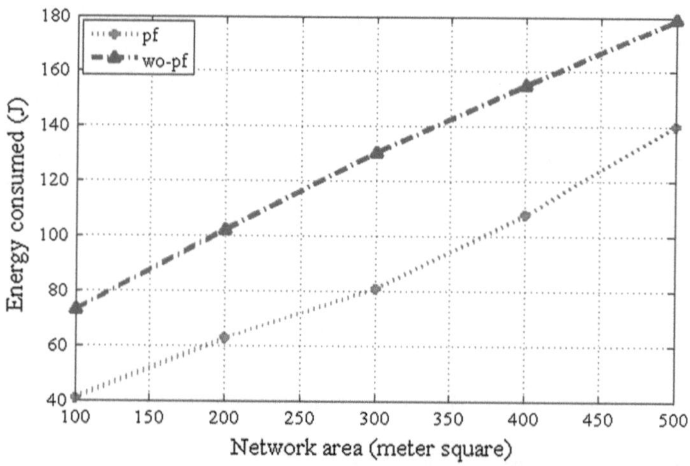

Fig. 17.2 Energy consumed with varying network area

We assume that the temperature of an agriculture land's surface is to be monitored continuously. The proposed algorithm is evaluated on the basis of average end-to-end delay, energy consumed, and network lifetime of the network.

This paper shows proposed filtering approach as pf and without proposed filtering approach as wo-pf. Figures 17.1, 17.2, and 17.3 shows the result of the proposed filtering approach by varying network size, number of cluster head nodes, and number of sensor nodes.

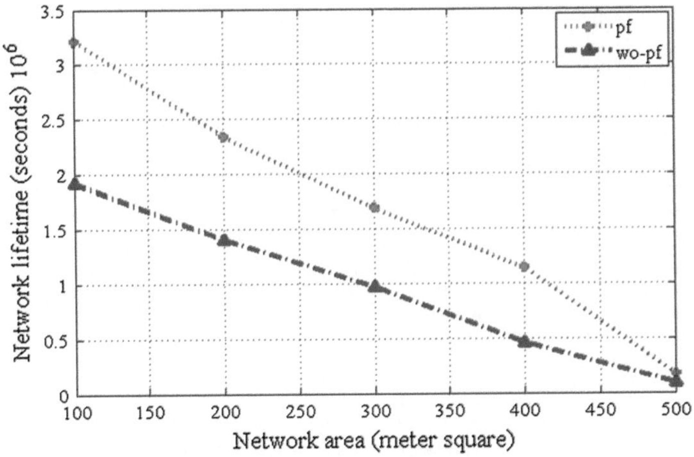

Fig. 17.3 Network lifetime with varying network area

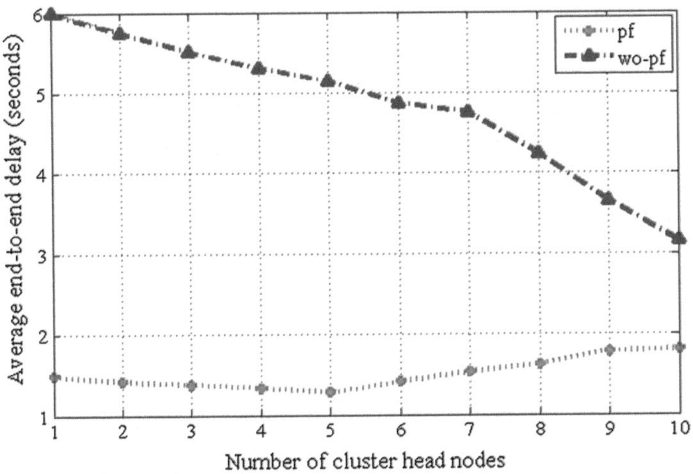

Fig. 17.4 An end-to-end delay with varying CH nodes

17.5.1 Varying the Network Size

The network area for the basic scenario was kept a square of $(100 \times 100)\text{m}^2$. The network size varied from $(100 \times 100)\text{m}^2$ to $(500 \times 500)\text{m}^2$ with an increase of $(100 \times 100)\text{m}^2$ every time. Variation in network size has a lot of impact on the network performance.

Figures 17.1 and 17.2 represent average end-to-end delay and energy consumption, respectively with varying network size. The figures shows that average end-to-end delay and energy consumption have increased because node density

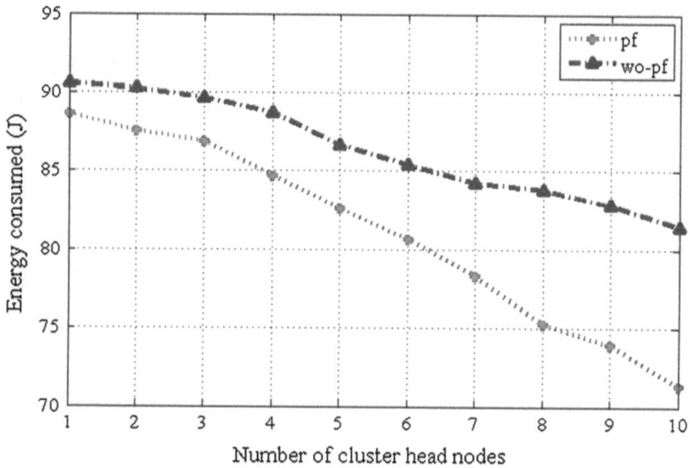

Fig. 17.5 Energy consumed with varying CH nodes

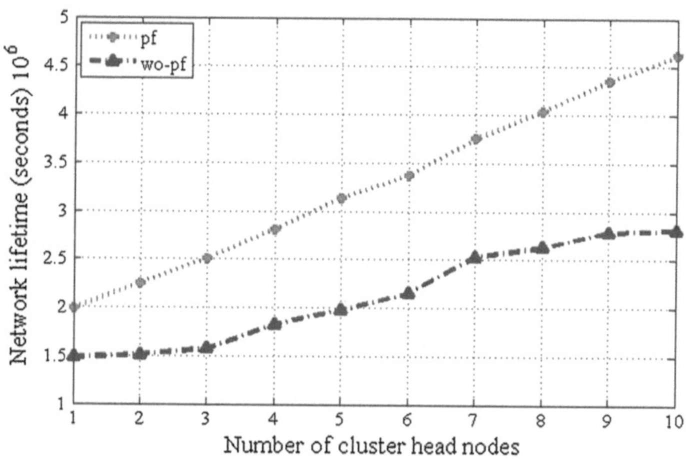

Fig. 17.6 Network lifetime with varying CH nodes

directly affects the network performance. The sparsely populated sensor nodes in a network form a very difficult environment because of poor connectivity.

Figure 17.3 represents network lifetime with varying network size. As shown in the figure, the network lifetime is deteriorating with increasing network area because the average path length between two communicating nodes is increasing and hence increasing the chances of link failure and hampers the repair mechanism.

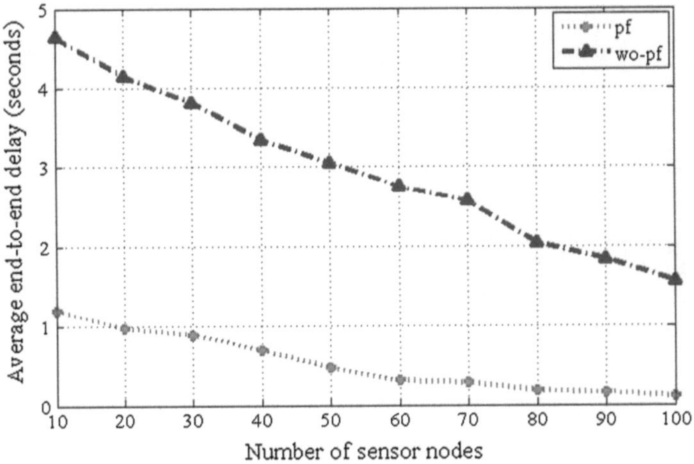

Fig. 17.7 An end-to-end delay with varying sensor nodes

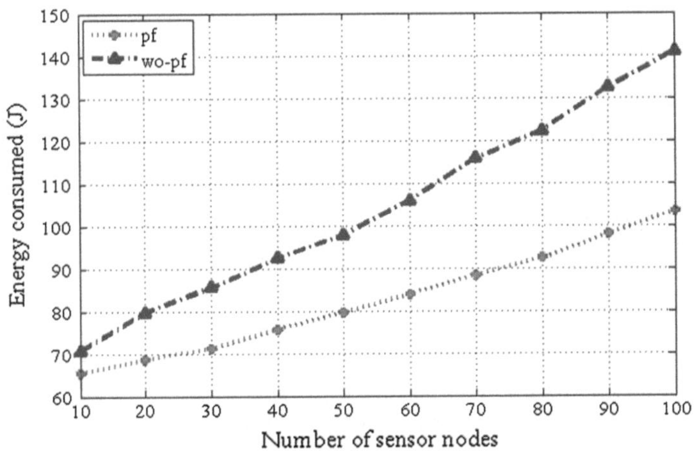

Fig. 17.8 Energy consumed with varying sensor nodes

17.5.2 Varying Number of Cluster Head Nodes

The number of cluster heads varied from 1 to 10 with an increase of 1 cluster head node every time during simulations. By increasing the number of cluster head nodes in the network, number of clusters increases and hence decreasing the cluster size.

Figures 17.4, 17.5, and 17.6 shows the results by varying number of cluster head nodes from 1 to 10. An average end-to-end delay in Fig. 17.4 and energy consumption in Fig. 17.5 is improving because of decreasing cluster size with

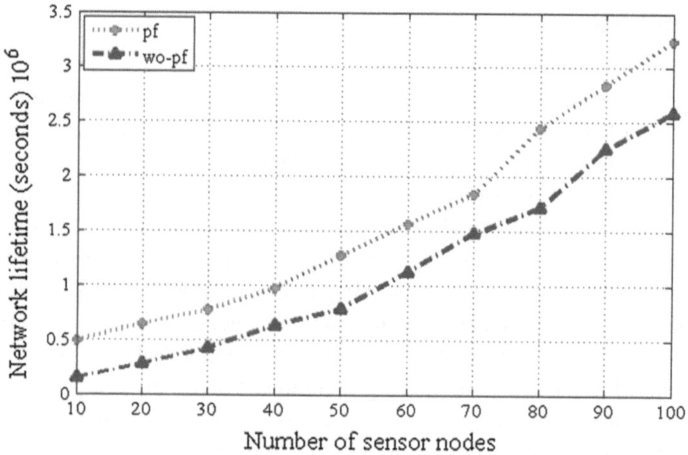

Fig. 17.9 Network lifetime with varying sensor nodes

increase in number of cluster head. Small-sized clusters can do better clustering and improves the network performance.

Network lifetime shown in Fig. 17.6 also improves because number of clusters in a network directly influences the network performance as well-clustered networks provide good connectivity of the sensor nodes.

17.5.3 Varying the Number of Sensor Nodes

The number of sensor nodes was kept as 100 for the basic scenario. In this section, number of sensor nodes varied from 10 to 100 with an increase of 10 sensor nodes every time.

Figures 17.7, 17.8, and 17.9 show the results by varying number of sensor nodes from 10 to 100. An average end-to-end delay in Fig. 17.7 is improving because node density directly influences the network performance as well distributed networks form a very good environment because of good connectivity of sensor nodes. Hence, more nodes of sensor nodes in an area having less inter-node transmission distance will reduce the average delay. The energy consumption in Fig. 17.8 is increasing because variation in number of sensor nodes increases the size of the network and hence increases the overall energy consumption.

Figure 17.9 shows improvement in network lifetime of the network because increase in density of nodes affects the average path length between two nodes that are communicating with each other. The chance of link failure reduces and network performs well.

17.6 Conclusions

This paper shows that the proposed filtering scheme outperforms the simulation when used with VAS. The VAS clustering approach is compared with already existing approaches and proved better than them. Hence, in this paper no other routing or clustering approach is taken into consideration. This paper is a step toward balancing the benefits of filtering along with clustering and routing in resource-limited environment. The proposed filtering technique outperforms in terms of all the above said parameters in varying conditions. Hence, our proposed approach shows promising results for improving network lifetime and energy efficiency of the network. Further research in this area will offer promising results. Various fusion techniques can also be used along with the filtering technique for in-network processing so that data communicated over the network can be further reduced. This filtering technique can be combined with any other clustering/routing algorithm for data filtering and improving energy efficiency and network lifetime of the network.

References

1. Abdelgawad, A., Bayoumi, M.: Resource aware data fusion algorithms for wireless sensor networks. LNEE, vol. 118, pp. 17–34, Springer, Heidelberg (2012)
2. Du, H., Hu, X., Jia, X.: Energy efficient routing and scheduling for real-time data aggregation in WSNs. J. Comput. Commun. **29**, 3527–3535 (2006) (Elsevier)
3. Zhou, B., Ngoh, L., Lee, B., Fu, C.: HDA: A hierarchical data aggregation scheme for sensor networks. J. Comput. Commun. **29**, 1292–1299 (2006) (Elsevier)
4. Luo, H., Luo J., Liu, Y., Das S.: Adaptive data fusion for energy efficient routing in wireless sensor networks. IEEE Trans. Comput. **55**(10), 1286–1299 (2006)
5. Verdone, R., Dardari, D., Mazzini, G.: Signal processing and data fusion techniques for WSANs. Wirelss Sens. Actuators Netw. (2008)
6. Zhu, Y., Vedantham, R., Park, S., Sivakumar, R.: A scalable correlation aware aggregation strategy for wireless sensor networks. J. Inf. Fusion **9**, 354–369 (2008) (Elsevier)
7. Hua, C., Yum, T.: Data aggregated maximum lifetime routing for wireless sensor networks. J. Ad Hoc Netw. **6**, 380–392 (2008) (Elsevier)
8. Luo, H., Tao, H., Ma, H., Das, S.K.: Data fusion with desired reliability in wireless sensor networks. IEEE Trans. Parallel Distrib. Syst. **22**(3), 501–513 (2011)
9. Xu, M., Leung, H.: A joint fusion, power allocation and delay optimization approach for wireless sensor networks. IEEE Sens. J. **11**(3), 737–744 (2011)
10. Renjith, P.N., Baburaj, E.: An analysis on data aggregation in wireless sensor networks. In: IEEE International Conference on Radar, Communication and Computing, pp. 62–71 (2012)
11. Zhang, H., Ma, H., Li, X.-Y., Tang, S.: In-network estimation with delay constraints in wireless sensor networks. IEEE Trans. Parallel Distrib. Syst. **24**(2), 368–380 (2013)
12. Ren, F., Zhang, J., He, T., Chen, C., Lin, C.: Attribute-aware data aggregation using potential-based dynamic routing in wireless sensor networks. IEEE Trans. Parallel Distrib. Syst. **24** (5), 881–892 (2013)
13. Cheng, C.T., Leung, H., Maupin, P.: A delay-aware network structure for wireless sensor networks with in-network data fusion. IEEE Sens. J. **13**(5), 1622–1631 (2013)
14. Lu, Z., Tan, S.-L., Biswas, J.: D2F: a routing protocol for distributed data fusion in wireless sensor networks. J. Wireless Pers. Commun. **70**, 391–410 (2013) (Springer)

15. Lu, Z., Tan, S.-L., Biswas, J.: Fusion function placement for active networks paradigm in wireless sensor networks. J. Wireless Netw. **19**(7), 1–12, 1525–1536 (2013) (Springer)
16. Khaleghi, B., Khamis, A., Karrey, F., Razavi, S.: Multisensor data fusion: a review of the state-of-art. J. Inf. Fusion **14**, 28–44 (2013) (Elsevier)
17. Sharma, T.P., Joshi, R.C., Misra, M.: Data filtering and dynamic sensing for continuous monitoring in wireless sensor networks. Int. J. Auton. Adapt. Commun. Syst. (IJAACS) **3**(3), 239–264 (2010) (Inderscience)
18. Gautam, N., Sofat, S., Vig, R.: Energy efficient voronoi ant system clustering algorithm for wireless sensor networks. In: EAI 5th International Conference on Adhoc Networks (ADHOCNETS 2013), Barcelona, Spain (2013)

Chapter 18
A Secure and Efficient Authentication Protocol in VANETs with Privacy Preservation

Chandra Sekhar Vorugunti and Mrudula Sarvabhatla

Abstract Vehicular Ad Hoc Network (VANET) is a kind of special mobile adhoc network. VANETs are used in many safely related services involving traffic management, route planning, and safety messaging (intelligent transportation systems), etc. through inter-vehicle and vehicle-to-infrastructure communications. VANETs have unique characteristics like high mobility of nodes and rapid changing topology which pose many challenging research issues in areas like data dissemination, data sharing, and security. The traditional security features are not suitable for VANETs. In 2011, Das et al. proposed a protocol based on hierarchical model for node authentication in group communications in VANETs and claimed that their protocol is robust against conventional security attacks. In this paper, we will show that Das et al. scheme cannot withstand to various conventional security attacks and fails to provide sender authentication which is important security requirement in VANETs. We then present our improved and generalized scheme (not specific to Das et al. scheme) to overcome the vulnerabilities stated in Das et al. scheme while preserving all the merits of their scheme.

Keywords Wireless communications • Vehicular adhoc networks (VANET) • Hierarchical model • Authentication • Security • Privacy • Confidentiality

---

C. S. Vorugunti (✉)
Dhirubhai Ambani Institute of Information and Communication Technology,
Gandhi Nagar, Gujarat 382007, India
e-mail: vorugunti_chandra_sekhar@daiict.ac.in

M. Sarvabhatla
Sri Venkateswara University, Tirupathi, Andhra Pradesh 517502, India
e-mail: mrudula.s911@gmail.com

R. Maringanti et al. (eds.), *Proceedings of Ninth International Conference on Wireless Communication and Sensor Networks*, Lecture Notes in Electrical Engineering 299,
DOI: 10.1007/978-81-322-1823-4_18,

18.1 Introduction

Road traffic activities are one of the most daily routines of common man. The increasing road accidents and traffic congestion are becoming major problems. VANET, a subset of Mobile Adhoc Networks is developed for this purpose which provides scalable and cost-effective solutions for applications such as safety messaging [1, 2] dynamic routing using DSRC [1, 3, 4] (Dedicated Short Range Communication). The appropriate integration of On-Board Unit (OBU) and positioning devices such as GPS receivers along with RSU and WAVES (Wireless Access in Vehicular Environments) opens tremendous research opportunities and challenges for the researchers [1, 2, 4–6]. As VANETs are used in much safety critical applications, one of the applications considering in this paper is secure safety messaging which is meant for cooperative driving and avoidance of accidents.

18.1.1 Security in VANETs

The security is critical in VANETs. It is essential to make sure that life-critical information cannot be tampered (inserting or modifying) by an adversary. The privacy and authenticity of the entities are to be maintained. Huang et al. [7] have shown that the security requirements in MANETs and other adhoc networks are not appropriate to be directly used to VANETs. Raya and Hubaux [8] have proposed first of its kind of security protocols for VANETs in systematic and quantifiable way. In 1991, Arazi et al. [9] proposed first of its kind of protocol for VANET communication using public key cryptography in the form of digital signatures which was accepted by lots of researchers [10–12].

In 2011, Das et al. [13] proposed an authentication protocol for VANET which uses hierarchical model. Das et al. [13] claimed that their scheme is robust to various conventional security attacks. We propose an improvement scheme over Das et al.'s [13] scheme to remedy their drawbacks while preserving all the merits of their schemes.

The rest of the paper is organized as follows. In Sects. 18.2, 18.3 a brief review of Das et al. [13] scheme is given. Section 18.4 describes the security weakness of Das et al. [13] scheme. In Sect. 18.5, our improved scheme is proposed and its security analysis is discussed in Sect. 18.6. The performance analysis of both the protocols is given in Sects. 18.7 and 18.8 provides the conclusion of the paper.

18.2 Review of Das et al. Protocol

In this section, we examine a cross-authentication protocol in VANET hierarchical model proposed by Das et al. [13] in 2011. Before going into the protocol, we explain the VANET hierarchical model and polynomial interpolation scheme on which the Das et al. [13] scheme is based on.

18.2.1 VANET Hierarchical Model (VHM)

To address the scalability and cross-certification issues, Das et al. [13] proposed a hierarchical protocol in which a complete hierarchy of CA is built which operates in a tree like structure to manage the network of vehicles. The VHM consists of two types of nodes: the powerful nodes and the leaf nodes. The powerful nodes are Certifying Authorities (CAs) as they considered being full of resources like memory and computation capability. The vehicles are light nodes having computation constraints. All the parents in the tree are CAs and the leaf nodes are vehicles. The root of the tree is Head CA (HCA) and next level of CAs is State level CA (SCA) which is nominated by HCA. Further SCA can have City CAs (CCA). The new level of CAs can be added according to the requirement of scalability. At leaf levels there will be no CAs but have vehicle nodes.

18.2.2 Polynomial Interpolation Scheme (PIS)

Based on Shamir [14] scheme, Dipanwitha [15] exploited polynomial interpolation scheme for multiparty authentication. Das et al. [13] scheme uses these PIS to frame the session key. As per this scheme there is a powerful node and light nodes similar to VANET heterogeneous environment. Let $U = \{U1, U2, U3, \ldots ,Un\}$ be the set of light nodes. Let 'M' be the powerful node. All the light nodes send their own contribution to powerful node with their IDs assumed by M. The powerful node "M" adds its own contribution and will have (ID_i, C_i) for each node U_i where ID_i be the identity and C_i is the contribution of the ith light node. Assuming each such pair from the "n" light nodes and one from "M", "M" will have totally $n + 1$ points. Assuming each such pair on the cartesian plane with the unique x-coordinates, IDs being unique, there passes only one "n" degree polynomial through the $n + 1$ points. The "n" degree polynomial will be of the following form: $a_0 + a_1x + a_2x^2 + a_3x^3 + \cdots + a_nx^n$. These $n + 1$ coefficients (a_0, a_1, a_2, $a_3 \ldots a_n$) when padded are used as session key, i.e., session key $= a_0 || a_1 || a_2 || a_3 || \ldots || a_n$. A brief introduction to polynomial Interpolation scheme is available in [16].

18.3 VH Protocol

VH Protocol proposed by Das et al. [13] consists of mainly three steps: first one is to initially set up the infrastructure, generating the session key between parent CA and child CA, and leaf level authentication (between two vehicles when they try to exchange a safety message).

18.3.1 Generating the Session Key Between Parent CA and Child CA

VH protocol requires setting up complete hierarchy of CA from top to bottom as discussed in Sect. 18.2.1. It is one-time infrastructure setup. The HCA will find trusted CA in state to take care of its own region. The public keys of all trusted CAs are known to other CAs. Those invited SCA are invited to send their contribution and their IDs are assumed by inviting CA (in this case CCA). The assumed IDs are not disclosed to the child CA. Based on the contributions received from the CCA, a session key is generated as discussed in Sect. 18.2.2 using polynomial interpolation scheme.

18.3.2 The Notations Used in Both the Schemes are Listed Below

H	A collision free hash function
U	Set of invited or vehicular nodes (based on context)
U_0	CA supervising the vehicular nodes
(pu_i, pr_i):	Public private key pair of the ith CA
x_i	Contribution of the *i*th CA
(pu_0, pr_0)	Public Private key pair of the supervising CA
ID_i:	Automatically detected ID of the vehicle U_i
N	Serial number of the vehicle which is given by CA, so that the vehicle is tractable by the CA and ignored by others
(PubKeyi, PriKeyi):	Public Private key pair of *i*th vehicle node
(PubKeyCA, PriKeyCA)	Public Private key pair of CA
$Cert_{ui}$	Certification of the vehicle U_i
SigPrCA	Digital signature using the private key of CA

18.3.3 The Session Key Setup Between the Head CA and Trusted CA

Step 1. $U_i \rightarrow U_0$: $Epu_0\ (x_i||T)$ where T is the time stamp. Every vehicle will send its own contribution to the powerful node, encrypting with the public key of U_0.

Step 2. U_0 decrypts all the contributions using its private key, i.e., $Dpr_0\ (x_i||T)$.

Step 3. U_0 assumes all the IDs of the invited CAs.

Step 4. Generates the session key 'K' with the contributions received as discussed in Sect. 18.2.2 with $n + 1$points.

Step 5. $U_0 \rightarrow U_i$: $E_{pui}\ (H(ID_i||K||ID_0))$ where ID_0 is the ID of the Head CA who is currently nominating the CAs. The ID_0 value sent to each U_i is taken as the identity of each nominated CA.

18.3.4 Leaf Level Authentication

Leaf level authentication works on the concept of polynomial interpolation scheme as discussed in Sect. 18.2.2. If any vehicle enters in the supervision area of any CA, it detects the vehicle and gets the ID of the vehicle.

Step 1. When a vehicle enters in the supervision area of any CA, CA detects the vehicle and gets the ID of the vehicle through an On-Board Unit installed in it. Then CA broadcasts a message which is encrypted with the ID of the vehicle. CA $\rightarrow$ Vehicle: $E_{\mathrm{ID}i}(\mathrm{ID}_i \| \mathrm{pu}_0 \| N)$ where the serial number N is only for supervising CA to identify the vehicle but other vehicles in the group just ignore the serial number.

Step 2. All the vehicular nodes at the leaf level send their contribution to the immediate CA, (U_0 in the following case) and U_0 generates the session keys as follows: $U_i \rightarrow U_0$: $\mathrm{Epu}_0(\mathrm{ID}_i \| x_i \| \mathrm{T})$ where T is the time stamp.

Step 3. U_0 collects and checks if the time stamp is in permissible range. Session key is built using polynomial generated (discussed in Sect. 18.2.2). i.e., computes session key $K = a_0 \| a_1 \| a_2 \| a_3 \| \ldots \| a_n$.

Step 4. U_0 calculates P_i for each U_i. $P_i = \mathrm{K} \oplus (\mathrm{ID}_i \| x_i)$

Step 5. All the P_i are concatenated and a secret is generated by U_0. The secret that is broadcasted, i.e., $U_0 \rightarrow U$: $P_1 \| P_2 \| P_3 \| \ldots \| P_n$.

Step 6. The U_i extracts his chunk automatically from the broadcasted secret and gets its P_i. U_i extracts the session key from P_i as follows: $P_i \oplus (\mathrm{ID}_i \| x_i) = K$. U_i knows his ID i.e., ID_i and contribution x_i. This new session 'K' is used to encrypt and decrypt the safety messages among the vehicles.

Step 7. $A \rightarrow B$: $E_K(M \| T \| N_A)$ where M: Message to be sent to B, T: Timestamp, NA: The sequence number which is automatically assigned to the vehicle A by the CA for identifying the vehicle. The received vehicle ignores N_A. As B also shares the same session key K, it decrypts the packet received using it and checks if the timestamp sent is in permissible range. If yes accept the packet else ignore it. All the message transmissions are broadcasts not a direct transmission. The intended recipient receives the packet remaining all discards the packet.

18.4 Analysis of Weakness of Das et al. Scheme

In this section, we will show that Das et al. [13] scheme fails to provide user authentication, nonrepudiation of origin of safety message, tampering message, message integrity and vulnerable to sybil attack, node impersonation attack, and timing attack [1, 2, 6, 17].

Two major drawbacks in Abhjith Das et al. scheme is that the group key 'K' sent by the supervisory CA (who is head of the vehicle nodes) is used as session key to share the safety message between two individual nodes, i.e., A and B. As the session key is known to every vehicle node in the group, any legitimate vehicle can decrypt the message. The second major drawback is that on entry of new vehicle, each vehicle needs to send their contribution and the key is generated based on the new contributions by the head CA using PIS scheme. This method of generating a new GK and broadcasting it every time a vehicle enters a group is extremely inefficient and can cause network bottleneck at the CA when numerous vehicles leave and enter the group at the same time, which might be a very common case for VANETs.

18.4.1 Failure of Providing Message Confidentiality

The message sent by *A* to *B*, i.e., $E_K(M||T||N_A)$ indicates that the sender belongs to the same group but does not provide the service of confidentiality from the insider attackers. An adversary from the group who possesses the session key can decrypt the message. Hence Das et al. [13] scheme fails to preserve message confidentiality.

18.4.2 Failure of Providing Entity Authentication

Consider a scenario in which vehicle '*A*' sends a safety message to '*B*,' i.e., $E_K(M||T||N_A)$ encrypting with the common group session key. Now '*C*' an adversary who is part of a group can receive the message. As session key '*K*' is known to everyone in the group, '*C*' can decrypt the message $D_K(M||T||N_A)$ and alter the identity N_A.'*C*' can send the altered safety message by replacing N_A with a random number N_R i.e., $E_K(M||T||N_R)$ to '*B*'. On receiving the message, '*B*' can decrypt the message and checks for the validity of *T*. If it is fine it will accept. In this scenario, it is not possible for '*B*' to check the authentication of the sender. Hence Das et al. [13] scheme fails to provide entity authentication.

18.4.3 Failure of Assuring Message Integrity

Consider a scenario in which vehicle '*A*' sends a safety message to '*B*,' i.e., $E_K(M||T||N_A)$ encrypting with the common group session key. Now '*C*' an adversary who is part of a group can receive the broadcasted message. As session key '*K*' is known to everyone in the group, '*C*' can decrypt the message $D_K(M||T||N_A)$ and can alter the message and broadcast an altered safety message $E_K(M^*||T||N_R)$. (Hence Das et al. [13] scheme also suffers from message tampering attack). On receiving the message, '*B*' decrypts the message and checks for the validity of T, if it is fine, it will accept. In this scenario, it is not possible for '*B*' to check the integrity of the messages. Research shows that the encryption stands alone cannot provide message integrity. The messages may be tampered by the adversary. Hence, Das et al. [13] scheme fails to provide message integrity. However as mentioned in [8], the system should guarantee that life-critical information cannot be modified by an attacker.

18.4.4 Sybil Attack

Vehicle '*A*' can illegitimately claims multiple false identities. '*A*' can simply create arbitrary new false identities and frame safety messages like $E_K(M_1||T||N_{R1})$, $E_K(M_2||T||N_{R2})$, $E_K(M_3||T||N_{R3})$, etc. According to Das et al. [13] scheme, the received vehicle node will focus on the message and ignores the serial number or identity. Hence N_{R1}, N_{R2} etc. are ignored by the receivers. So in Das et al. [13] scheme a single node can create pseudonymous identities, and can bypass the reputation system and consequently cause DoS to legitimate node.

18.4.5 Node Impersonation Attack

As discussed in Sect. 18.4.3, a legal vehicle '*A*' can illegitimately impersonate some nonexistent nodes and creating safety messages like $\underline{E_K}(M_1||T||N_{R1})$, $E_K(M_2||T||N_{R2})$, $E_K(M_3||T||N_{R3})$, etc. No unique identity of sender is needed to frame the safety messages, so any legitimate user having the session key '*K*' can frame valid messages to circulate in the group. Hence Das et al. [13] scheme suffers from node impersonate attack.

18.4.6 Timing Attack

Consider a scenario in which a vehicle '*A*' sends a safety message to '*B*', i.e., $E_K(M||T||N_A)$ encrypting with the common group session key. Now '*C*', an adversary who is part of a group, receives the message. As session key '*K*' is known to everyone in the group, '*C*' can decrypt the message $D_K(M||T||N_A)$ and can add delay into the message $E_K(M||T + \Delta t||N_R)$ without altering any content of the message and these messages are received after the permissible time range by the intended recipient. Safety messages are very time critical. If delay occurred in these messages, then the ultimate purpose of the protocol is not achieved. Hence, the Das et al. [13] scheme suffers from the biggest drawback of timing attack which makes the complete message useless.

18.5 Our Proposed Authentication Protocol

In our proposed protocol, we use Public Key Infrastructure (PKI). Each vehicle '*i*' is assigned with a set of public and private key pairs (PubKeyi, PriKeyi). The supervising CA can issue and sign these public keys. The public key of the CA is dynamically transmitted to all the vehicles. The vehicles store these public and private key pairs in tamper-proof hardware which will keep the information safe from attackers. A certificate of public key of a vehicle V should contain CertV[PubKeyV] = PubKeyV || $Sig_{PriKeyCA}$ [PubKeyV||ID_{CA}]. In our protocol two types of session keys are considered, pairwise (between two communicating vehicles) and group key (for entire group broadcast). In our protocol, session key encrypts the message in addition to it the encrypted message is also signed by the sender private key which provides confidentiality and nonrepudiation.

18.5.1 Establishing Group Key Between Vehicles and Supervising CA

Step 1. CA → U_i: E_{ID_i}(PubKeyCA, PubKey*i*, PriKeyi).CertU_i. When a vehicular node U_i enters into the supervision area of any particular CA, On-Board Unit of that vehicle transmists a radio signal which inidiates the entry of U_i into the supervision area of CA. CA assigns an identity i.e., ID_i to U_i and sends it to On Board Unit of U_i by secure control channel. The supervisor CA forwards a message which contains public key of CA, public key and private key pair assigned to U_i which is encrypted with ID_i. Supervisor CA also forwards public key certification of U_i i.e., $Cert_{Ui}$ = {PubKeyi||$Sig_{PriKeyCA}$ [PubKeyi ||ID_{CA}]}.

Step 2. CA calculates the Group Key GK (picks up any random number) of length 128 bits.

Step 3. CA → U_i: Message = *H*(PubKeyA), $E_{PubKeyA}$(GK), *H*(PubKeyB), $E_{PubKeyB}$(GK) etc., $Sig_{PriKeyCA}$\{Message\} (where *A*, *B* are vehicles in the group). In our scheme, the supervisor CA calculates group key without using polynomial interpolation scheme. Once CA framed the group key, it broadcasts to all the vehicles under his supervision. CA will sign the message with its private key. Group key generation is one-time operation performed by the CA.

Step 4. When a new vehicle enters the group, CA sends a broadcast signal containing a random number '*n*', which informs the vehicle nodes to apply '*n*' one bit cyclic shift on current GK instead of executing the complete key generation algorithm (which is the drawback in Das et al. scheme). Along with vehicle nodes, the CA also performs '*n*' one cyclic shift on the current GK on entering new vehicle. If new vehicle enters, CA sends the current GK of the group to the new vehicle as follows: *H*(PubKeyN), $E_{PubKeyN}$(GK), SigPr iKeyCA{$E_{PubKeyN}$\{GK\}, *H*(PubKeyN)}. Once the group key is established, if two vehicles *A* and *B* wants to exchange safety messages (not a broadcast), a pairwise session key is established between them.

18.5.2 Establishing Session Key between Vehicles A and B

Two vehicles *A* and *B* which belong to same group and want to communicate; a pairwise session key is established between them.

Step 1. $A \rightarrow B$: $M1$, $T1$, $\mathrm{Sig}_{\mathrm{PriKeyA}}[M1 \| T1 \| \mathrm{GK}]$, CertA.
A sends a message to initiate the safety message communication with *B*, it signs the message with its private key. The message contains invitation message *M*1, Time stamp, Group Key, and CertA. The GK indicates to *B* that *A* belongs to same group (not an outsider). Through CertA, B can cross-check the authentication of *A*. Once *B* authenticates *A*, it will responds to the request. *A* belongs to same group and the message is signed by sender private key which provides both senders authentication and nonrepudiation.

Step 2. $B \rightarrow A$: $M2$, $T2$, $\mathrm{Sig}_{\mathrm{PriKeyB}}[M2 \| T2 \| \mathrm{GK}]$, CertB
Similar to *A*, *B* also sends the reply message *M*2 and a message signed by its private key which contains *M*2, Time stamp and group key. *B* also forwards CertB.

Step 3. '*A*' constructs the session key SK.

Step 4. $A \rightarrow B$: $E_{\mathrm{PubKeyB}}\{M3 \| \mathrm{SK} \| T3 \| \mathrm{GK}\}$, $\mathrm{Sig}_{\mathrm{PriKeyA}}\{\mathrm{MAC}(M3 \| T3 \| \mathrm{GK}, \mathrm{SK})\}$
Once session key SK is framed, '*A*' frames a message which is encrypted with the public key of *B* which contains the message *M*3, the session key SK, Time stamp, Group Key, and MAC of the contents also signed by the private key of *A*, which assure authenticity and nonrepudiation.

Step 5. $A \rightarrow B$: $E_{\mathrm{SK}}[M4 \| T4]$, $\{\mathrm{MAC}(E_{\mathrm{SK}}(M4 \| T4), \mathrm{SK})\}$
Now once *A* and *B* shared the session key securely with mutual authentication, *A* sends a message to *B* which is encrypted with session key shared with *B* and also the MAC value of the message *M*4 (Safety message) with the session key which helps for *B* to check for the message integrity (which is not available in Das et al. scheme). As the session key is exchanged in a secure manner between *A* and *B*, once it receives the message encrypted with the SK, *B* is assured that the message is from *A* only which provides nonrepudiation. Even if the message is taken by any adversary in the group, he does not know the session key SK shared between *A* and *B*. Hence it is not possible to decrypt the message *M*4. Usage of MAC in step 5 also ensures the authenticity of the sender and message integrity. The encrypted function can be selected from symmetric encryption functions such as AES.

18.6 Security Analysis

In this section, we analyze the security features provided by our scheme.

18.6.1 Message Confidentiality

In our scheme, the session key is shared in very secure manner by the message $E_{\text{PubKeyB}}\{M3||\text{SK}||T3||\text{GK}\}$. The message is encrypted with the public key of *B*, so that the decryption must be done with private key of *B* only. Hence it is not possible to any kind of user to get the SK shared between *A* and *B*. Once the session key is shared, the safety messages are exchanged by encrypting the message with the session key $E_{\text{SK}}[M4||T4]$. Therefore, no one other than *B* can able to decrypt the message. Hence, the confidentiality of session key and safety message is assured in our scheme.

18.6.2 Entity Authentication

In our scheme, the receiver can check the authenticity of the sender in two ways. First, while initializing the request, *A* sends the message by signing with his private key which must be decrypted only with the public key of *A*. Along with the initiation message, *A* also sends the certification (CertA) issues by CA. From the certification sent by *A*, *B* will get the public key of *A* and decrypts the message. Hence *A* is authenticated. While exchanging the message in session, A encrypts the message with SK which is shared between *A* and *B* only. Hence *B* ensures that the message is coming from *A* only. So in our scheme the receiver can authenticate the sender in two ways.

18.6.3 Message Integrity

In Das et al. [13] scheme there is no provision for the receiver to check the integrity of the messages (No padding of digested code like MAC, etc.). In our scheme, *A* along with the safety message sends the message $\text{Sig}_{\text{PriKeyA}}\{\text{MAC}(E_{\text{SK}}(M4), \text{SK})\}$. Signing the message with the private key of *A* ensures that the message is from *A* only. Even if the message is taken by any adversary in the group, he does not know the session key SK shared between *A* and *B*. Hence it is not possible to decrypt the message *M*4. Hence our scheme assures message integrity.

18.6.4 Resisting Sybil Attack

In Sybil attack, Vehicle '*A*' can illegitimately claims multiple false identities. 'A' can simply create arbitrary new false identities and frame safety messages. In our scheme, even A creates false identities but it is not possible for *A* to create their certificates. The supervisor CA will issue public key certificates to the group members. Without providing the valid certificate issued by supervision CA, no intended receiver will respond to the invitation message. In the first place, it is not possible for *A* to create fake certificates. Hence in our scheme Sybil attack is not possible.

18.6.5 Resistance to Node Impersonation Attack

If vehicular node *A* tries to impersonate as any other legal node, it must posses the certificate issued by the CA for that node. The certificates are issued by only supervision CA and the certificates are CertV = VPubKey|| $\text{Sig}_{\text{PriKeyCA}}$[VPubKey||$\text{ID}_{\text{CA}}$] are digitally signed by the CA with his private key. The private key of CA is not known to anyone. Hence even if a vehicular node illegitimately impersonates some nonexistent nodes and creates safety messages, it is not possible to frame a certificate which the recipient trusts. Hence our scheme is secure against node impersonation attack.

18.7 Performance Evaluation

In this section, we compare performance and security features of proposed scheme with Das et al. [13] scheme. The computation costs required by the leaf level nodes are considered. The comparison results are depicted in Tables 18.1 and 18.2. The first two columns of Table 18.1 compares the cost required in the initial group key and session key setting stage. The third and fourth column compares the cost required when a new vehicle enters the group.

From Table 18.1, we can conclude that the proposed scheme requires very less number of cryptographic operations in the initial key generation phase itself. Once a new vehicle enters, the Das et al. [13] scheme executes complete key generation algorithm, whereas our scheme executes only cyclic bit shift which requires very negligible operations. The major operation which requires more computation in the proposed protocol is digital signature. Fast and efficient digital signature implementations are available [10, 18]. Hence we can use digital signatures which require less computation. In Das et al. [13] scheme, the supervisor CA uses polynomial interpolation scheme to generate a session key which involves complex algebraic and calculus operations. Compared to the computation cost required for digital signatures, PIS requires $O(n^2)$ operations [19] where 'n' is the degree of the polynomial (no. of vehicles which submitted their contribution to the supervision CA).

Table 18.1 Cost comparison in the initial group key and session key formation

Cryptographic operations	Das et al. scheme	Proposed scheme	Das et al. scheme	Proposed scheme
Encryption	$n+1$	$0+2$	n	0
Decryption	$n+1$	$n+2$	n	1
XOR	n	Nil	n	Nil
Hash	Nil	n	Nil	Nil
Mac	Nil	4	Nil	0
Digital Sign	Nil	$n+6$	Nil	1
Polynomial Interpolation	$O(\mathrm{n}^2)$	Nil	$O(n^2)$	Nil

Table 18.2 Comparison of security features

Security feature	Das et al. scheme	Proposed scheme
Provides message confidentiality	No	Yes
Provides entity authentication	No	Yes
Provides message integrity	No	Yes
Withstand sybil attack	No	Yes
Withstand node impersonation attack	No	Yes
Withstand timing attack	No	Yes
Providing nonrepudiation	No	Yes
Withstand DOS attack	No	Yes

18.8 Conclusion

More recently, Das et al. proposed an efficient cross-authentication protocol in VANET hierarchical model and claimed that their protocol satisfies major security features. However, in this paper we have shown that Das et al. scheme fails to provide major security features like entity authentication, message integrity, etc. and fails to resist various attacks like timing attack, Sybil attack, node impersonation attack, etc. As part of our contribution, we have proposed an improved and efficient authentication protocol to remedy the security flaws in Das et al. scheme.

References

1. Xu, Q., Mak, T., Ko, J., Sengupta, R.: Vehicle-to-vehicle safety messaging in DSRC. In: Proceedings of the 1st ACM International Workshop on Vehicular Ad Hoc Networks, Philadelphia, PA, USA, 01 Oct 2004
2. Yang, X., Liu, J., Zhao, F., Vaidya, N.: A vehicle-to-vehicle communication protocol for cooperative collision warning. In: First Annual International Conference on Mobile and Ubiquitous Systems: Networking and Services (MobiQuitous 2004) 2004
3. 5.9 GHz DSRC, http:grouper.ieee.org/groups/scc32/dsrc

4. Duri, S., Gruteser, M., Liu, X., Moskowitz, P., Perez, R., Singh, M., Tang, J.-M.: Framework for security and privacy in automotive telematics. In: Proceedings of the 2nd International Workshop on Mobile Commerce, Atlanta, Georgia, USA, 28 Sept 2002
5. Jakubiak, J., Koucheryavy, Y.: State of the art and research challenges for VANETs. In: Consumer Communications and Networking Conference, 5th IEEE, 10–12 Jan 2008, pp. 912–916
6. Mishra, B., Saroj Kumar, B., Jena, D., Jena, S.K.: A secure and efficient message authentication protocol for VANETs with privacy preservation. In: World Congress on Information and Communication Technologies, pp. 884–889, Dec 2011
7. Huang, Y.-M., Hsieh, M.-Y., Wang, M.-S.: Reliable transmission of multimedia streaming using a connection prediction scheme in cluster-based ad hoc networks. Comput. Commun. **30**(2), 440–452 (2007)
8. Raya M, Hubaux, J.-P.: The security of vehicular ad hoc networks. In: Proceedings of the 3rd ACM Workshop on Security of Ad Hoc and Sensor Networks, Alexandria, VA, USA [doi.10.1145/1102219.1102223], 07 Nov 2005
9. Arazi, B.: Vehicular implementations of public-key cryptographic techniques. IEEE Trans. Veh. Technol. **40**, 646 (1991)
10. Cramer, R., Shoup, V.: Signature schemes based on the strong RSA assumption. IBM Research Report RZ 3083, Institute for Theoretical Computer Science, ETH Zurich, 8092 Zurich, Switzerland
11. Furgel, I., Lemke, K.: A review of the digital tachograph system, In: Proceedings of the Workshop on Embedded Security in Cars (escar)'04, 2004
12. Boyd, C., Mathuria, A.: Protocols for Authentication and Key Establishment. Springer, Berlin (2003)
13. Das, A., Chowdary, D.R., Rai, A.: An efficient cross authentication protocol in VANET hierarchical model. Int. J. Mob. Adhoc Netw. **1**(1), 128–136 (2011)
14. Shamir, A.: How to share a secret. CACM **22**, 612–613 (1979)
15. Dipanwita, R.C., Mounita, S.: An efficient group key agreement protocol for heterogeneous environment. In: SECRYPT'09
16. Wang, Q., Moin, P., Iaccarino, G.: A rational interpolation scheme with superpolynomial rate of convergence. SIAM J. Num. Anal. **47**(6), 4073–4097 (2010)
17. Hubaux, J.-P., Čapkun, Srdjan, Luo, J.: The security and privacy of smart vehicles. IEEE Secur. Priv. **2**(3), 49–55 (2004)
18. Gollan, L., Meinel, C.: Digital signatures for automobiles. Proc of Systemics, Cybernetics and Informatics (SCI), (2002)
19. Rajasekaran, S., Reif, J.: Handbook of Parallel Computing: Models, Algorithms and Applications. Chapman & Hall/CRC, London (2007)

18. [illegible] (2004)

19. [illegible] (2008)

20. [illegible]

21. [illegible] (2001)

Chapter 19
Design Approach of Self-Organized Routing Protocol in Wireless Sensor Networks Using Biologically Inspired Methods

A. N. Thakare and L. G. Malik

Abstract Wireless sensor networks are composed of a large number of nodes equipped with radios for wireless communication, sensors for sensing the environment, and CPU's for processing applications and protocols. A significant number of wireless sensor networks consist of battery-powered nodes to be able to operate unattended. Such networks require autonomy of management (self-organization), robustness, scalability, fault tolerance, and energy efficiency in all aspects of their operation. These properties are especially important for routing, since multi-hop communication is a primitive wireless sensor network operation that is robust, scalable, and adaptive with fault-prone as well as energy intensive. The objective is to design the routing protocol for robustness in self-organization in wireless sensor networks. In this paper, we try to design the novel architecture of robustness in self-organization with the consideration of three different bioinspired methods, i.e., BeeSensor, self-organized data gathering scheme (SDG), and AntHocnet for comparative study.

Keywords Wireless sensor networks • Ant colony optimizations • Bee colony optimizations • Routing protocols • Self-organization

19.1 Introduction

A wireless sensor network is a need for today's real-time applications as part of networks. The WSN is expected to remain operational for an extended period of time. During this time, new node may be added to the network, while other nodes

A. N. Thakare (✉) · L. G. Malik
G. H. Raisoni College of Engineering, Nagpur, India
e-mail: amu_thak@rediffmail.com

L. G. Malik
e-mail: latesh.malik@raisoni.net

R. Maringanti et al. (eds.), *Proceedings of Ninth International Conference on Wireless Communication and Sensor Networks*, Lecture Notes in Electrical Engineering 299,
DOI: 10.1007/978-81-322-1823-4_19,

Fig. 19.1 Self-organization in research field [19]

might incur in failures or exhaust their batteries, becoming unoperational. A routing protocol must be resilient to such dynamic and generally unpredictable variations and must sustain the availability of essential network. Therefore, network protocols and routing protocols must be empowered with self-organizing and self-management properties in order to make the network functioning an autonomic system (Fig. 19.1).

19.1.1 Functions as of Self-Organizing

- Self-organizing function in wireless sensor networks by following methods.
- Sharing Resources (processing and communication capacity).
- Forming and maintaining structures.
- Adopting behavior associated with routing.
- Assigning the task and configuring the software components with disseminating and queuing information.
- Managing Resources (synchronize time and conserving power).
- Provide robustness by repairing the faults and attack.

The self-organizing systems completely rely on localized decision processes which have the three aspects of successfulness of self-organization. First, positive and negative feedback; second, interactions among individuals with environment; and third, probabilistic techniques. By concentrating radio interferences required to develop the controlled self-organization by initializing with MAC layer. The controlled self-organization is important for the realization of large-scale wireless sensor networks associated with robustness to network topology changes is also important for wireless sensor networks.

The communication protocols are not sufficiently flexible regarding environmental changes. These environmental changes and control on each layer in wireless sensor network architecture operates on widely different timescales. MAC layer supports for one hops communication where data transmission takes few milliseconds in most sensor networks. Energy efficiency MAC protocol with sleep scheduling for prolonging lifetime are assumed in sensor networks. Whereas routing layers have to deal with topological changes to realize source to destination communications. In static sensors, nodes manage network topology by using HELLO message every tens of seconds. The external timescale operation control of self-organization is longer than routing layer. It is insufficient to discuss about the robustness within one layer. Here, we try to design the self-organized based routing protocols for the condition of channel selection, mobility, and node failure.

In MAC layer, the sleep control is expected so that power saving option is successful. The MAC protocol with the sleep control allows the node to sleep for every 10 ms. So that each node can communicate with other nodes only when it is awake. The cycle of sleep control means minimum unit time of one-hop data transmission. In MAC layer, during the selection of a next hope node when a node is in sleep mode, the data is held for a certain period of time. There is a condition for probabilistic channel selection for communication. The channel selection from spectrum sensing/sharing utilizing the available spectrum band is called as cognitive radios. In Nature, self-organization for sensor node in wireless sensor networks which is to be analogous to biologically inspired methods, i.e., availability of different paths for ANTs and BEEs for searching the food among different paths to reach to the destination i.e., sharing resources. Consider the insects' colony as a cognitive radio network and the insects as a cognitive radio for spectrum utilization. Task allocation is available channel and task associated stimuli as a permissible power to channel.

(a) Selecting the channel that has the minimum channel selection probability (appropriate channels) and avoiding the interference at the same time (select the paths by ants and bees as availability of different paths).

In the proposed protocol, we consider that decentralized architecture needs a powerful autonomous entity that reflects a behavior of adaptation. Here we consider the two main processes that are required (1) Signaling process where the routing tables are created and updated. (2) Routing process where the data packets are forwarded to sink node. The goal of the signaling process is to create pheromone gradient toward the sink. To achieve the goal, the forwarding attitude to the sink node is by emission and perception of pheromone signals and it is decreased by the pheromone evaporation in biological process. To achieve the signaling process the following tasks has to be performed. (1) Forwarding attitude estimation (2) Pheromone emission (3) Pheromone perception and (4) Pheromone evaporation. This is all about the signaling approach of more pheromone, i.e., more times path referred creates pheromone value high and less times referred pheromone value is less as shown in figure below. The node forwards the data packets to sink node with more pheromone value which has the concept of Positive feedback and Negative feedback. The positive feedback is the most powerful mechanism for building structure and negative mechanism that keep the process controllable. Here we consider the event detection in which the node decides whether or not to join the events in order to pheromone accumulated at that events wait for certain duration of time and leave the events. As the pheromone accumulates, the events

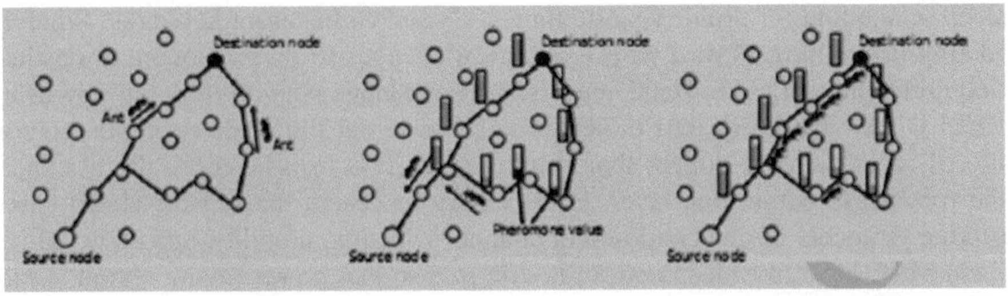

Fig. 19.2 The ACO-based routing process with increasing pheromone values

become higher and more nodes are attracted to join the events until enough nodes are there to complete the job of food searching. Deeper the color means higher pheromone level. The proposed method is more appropriate for the process of automaticity or self-organization. Here we consider the cluster-based data gathering with multi-sink option for the node to node communication within the cluster called cluster pheromone (Fig. 19.2).

(b) Manage the number of sensor nodes which are nearer to each other as a neighboring node forming the groups called cluster (the node that has the fewer radiuses). Gathering the data as a response to sink node for HELLO message as a multi-hop communication and accomplish the task of join regarding to pheromone levels.

The next process is the routing process. Here we consider the distributed routing which is more robust to network variation. In wireless sensor networks, node selects the next hop to communicate and sends it to the sink node and if the number of nodes is more than (multi-hop) the selection of the nodes, it forwards it to the sink node. In the routing process, if the large sensor nodes are deployed in WSN, it is necessary to create the Clustering approach which is to be introduced here. For self-organization, the uses of clustering approach are data gathering and attaining self-recovery capabilities. Some species in the world of ants divide their eggs and larvae based on their size. The clustering algorithm called ant-based clustering [1]. The architecture of self-organization in the topology changes provides the knowledge of self-organization. As shown in Fig. 19.3, the situations of node failure still the other node or other path is utilized for forwarding the data to the sink node. Though the topology changes still the sensor network is self-organized.

(c) If sink node fails to send the data to the base station. To improve the problem at that moment the multi-sink operation is useful.

Ant-based clustering is a probabilistic approach where clustering is repeatedly realized by ants and stochastically selected eggs are picked up or dropped (join and leave operation of sensor nodes). We introduced such a probabilistic approach to the clustering of sensor nodes. Dynamic clustering adapting to changes in the environment can accommodate sink node failures. The solution is using the multi-sink option.

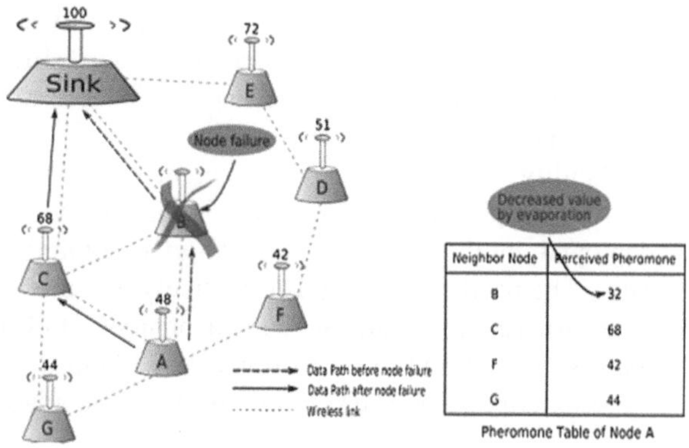

Fig. 19.3 The architecture of self-organized in topology

19.2 Literature Review

In wireless sensor networks, many reviews have been done on biological basis since 1990s onward. Most of the surveys are based on classical algorithms especially based on network layer. But here we concentrate on biological method based algorithms.

In wireless sensor networks [2] surveyed the classical and swam intelligence routing. The classical routing was surveyed in 2002 by Akyildiz et al. [3], Karaka and Kamal [4], Akkaya and Younis [5]. Whereas analogous to the property of classical routing, the social insects' communities survey, called as swarm intelligence (biologically inspired routing protocols), identified by Macro Darigo and Gi Caro [6] is a novel field originally defined as "any-attempt to design algorithms or distributed problems—solving devices inspired by collective behavior of social insects and animal societies."

The inspiration from these concepts the different engineering concepts are surveyed providing the distributed system that has different challenges observed that can show the adaptive fault tolerance, robust in self-organization and scalable routing behavior based on biologically inspired methods is ant colonies [6, 7], slimemold [8], particle swarm optimization [9], honey bees [10], flocks of birds, and schools of fishes.

19.3 Design Issues and Routing Factors in WSNs

A large number of researches have been carried out to overcome the constraints of WSN's and solve the design and applications issues. The challenges and design issues in WSN are power consumptions, fault tolerance, scalability, productive cost, quality of services, data aggregations and fusion, node mobility, connectivity, security, congestion, self-organization, and latency.

1. Self-Organization: A WSN is expected to remain operational for an extended period of time. During this time new nodes might be added to the network,

while other nodes might occur in failures or exhaust their batteries and become unoperational. A routing protocol must be resilient to such dynamic and generally unpredictable variations and sustainable to long-term availability of network services.
2. Scalability: Sensor nodes deployment in WSNs is application dependent and affects the performance of the routing protocol. Large number of nodes are deployed in the region having short communication range and high failure rates. The routing protocol is effectively acceptable to these challenges.
3. Fault Tolerance: If one of the sensor nodes fail, the algorithm can reorganize itself so that it continues to function without any disruption. The nodes are in working condition. The routing protocols are adaptive to topological changes.

19.4 Taxonomy of Routing Protocols

The taxonomy of routing protocols was surveyed and find out the some suitable architectures. According to many different authors' review the taxonomy of routing protocols. According to Akkaya and Younis, group routing protocols for WSNs are categorised into four different forms (a) data-centric (b) hierarchical based (c) Location-based, and (d) QOS-aware, whereas data-centric protocols do not require a globally unique ID for each sensor node and perform multi-hop routing. For the development of new routing protocols, we concentrate on biologically inspired methods as discussed below:

19.4.1 Biological Methods

The studies of social behavior of organisms (individuals) in swarm prompted the design of very efficient optimization and algorithms. The state of art of many algorithms are based on ant colony optimizations (ACO) and bee colony optimizations (BCO) techniques useful to solve combinatorial optimization and NP-hard problems.

19.4.2 Ant Colony Optimization

It is well known that the biological ants in real world are able to utilize swarm intelligence to find the solutions for different problems. ACO have been developed to mimic the behavior of real ants to provide heuristics solutions for optimization problems. It was first proposed by Darigo [11] in his Ph.D. dissertation when searching for food biologically ants exhibit complex social behavior based on the hormones they deposited are called pheromones. Pheromones attract other

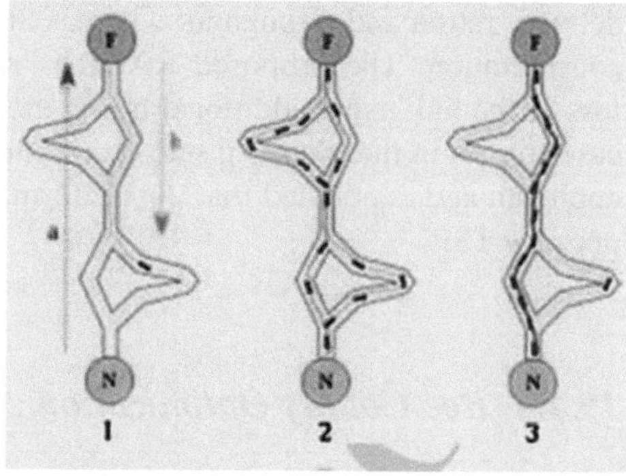

Fig. 19.4 Ant colony optimization

ants and outline a path to the food source that other ants can follow. Remaining other ants walk along the path, more pheromone is laid and path will increase and required time is less. There is another possibility of pheromone evaporation for that purpose reduces the change for other ants to take path (Michael Brand et al. 2010). This characteristic of ants is adapted on ACO algorithms to solve real problems. ACO meta-heuristics approached models are defined. In ACO a no. of artificial ants build solutions to optimization problems. The path optimization between nest and food is achieved by ants' colonies by exploiting the pheromone quantity dropped by ants. In WSN the path is chosen and data are transmitted through the labels on head of data packets, whereas in ACO the path is chosen based on pheromone left by ants in the path. Whereas there are forward ants and backward ants which have the responsibilities of forwarding ants to exploring paths and collect the information from source to destination and backward ants to updating the information and pass to the other nodes [12].

Figure 19.4 shows the behavior of Ant colonies. The first ant finds the food source (*F*), via any way (*a*), then returns to the nest (*N*), leaving behind a trail pheromone (*b*) ants follow four possible ways, but the strengthening of the runway makes it more attractive as the shortest route. Ants take the shortest route as long portions of other ways lose their trail pheromones. These inspire the routing techniques.

19.4.3 Overview of ACO-Based Algorithms

In network routing, Ant Net routing using ACO techniques provides better results. Comparing all routing algorithms with ACO provides that ants are small and can be piggybacked in data packets and the frequent transmission of ants may be possible in order to update the information to solve link failure. An ACO algorithm, which aims at minimizing the complexity in the nodes, is optimal for a less number of nodes in the cluster but not suitable for ad hoc networks. The fault-tolerant routing protocols using greedy ACO choosing a single path achieve a high packet delivery ration and throughput whereas the packet loss on the link is not taken into consideration. The proposed RACO is reconstructed for considering the packet loss of the link using additional traffic model. The performance of ACO on various case studies in the traveling salesman problem (TSP) is analyzed using a two-stage approach and concluded that the performance of ACO is optimal than the existing ones for TSP.

19.4.4 Bee Colony Optimization

The bee is a social and domestic insect native to Europe and Africa. There are between 60,000 and 80,000 living elements in a hive. The bees feed on a nectar as a source of energy for their lives and use pollen as a source of protein in the rearing larvae known as Queen Males and Drones. A several thousands of sterile females are called workers and many young bee larvae are called Broods. The basic working of BCO is another technique of swarm intelligence in which the social insects, i.e., bees are based on a few relatively simple rules of individual insects' behavior which is discovered by Lucic and Teodorovic [13] who first used these basic principles of collective bee intelligence in solving the combinational optimization problems. He introduced the bee system and tested in case of TSP. The main advantage of bees is self-organization using BCO. Many self-organization algorithms have been developed until now on various applications. The studies are mainly based on dance and communications, task allocation, collective decision, nest site selection, nest site selection, mating marriage, reproduction foraging, floral and pheromone layering, and navigation behaviors of the swarm. Some known algorithms based on bee SI are virtual bee, BeeAdhoc, the marriage in honey bees, the BeeHive, bee system, BCO, and ABC.

19.4.4.1 Bee Communication

The communication language of bees is of extreme precision based on dances. Bee scouts after finding food return to hive; this type of worker "scouts" inform others about distances and directions of food sources. If the dance is faster, then the food distance is smaller. The accuracy of direction (angle between the food source and the sun relative to hive) is within 30°. The amount of food depends on the wriggling of the bee. The more the wriggling is, the more the quantity is. The bee has different types of behavior such as foraging behavior in which nest site searching, food source searching, and marriage behavior. In this way a bee starts communication. The BCO is utilized for solving problems on different engineering aspects. The Self-organized routing approach in both classical-based and swarm-based routing protocols are discussed below.

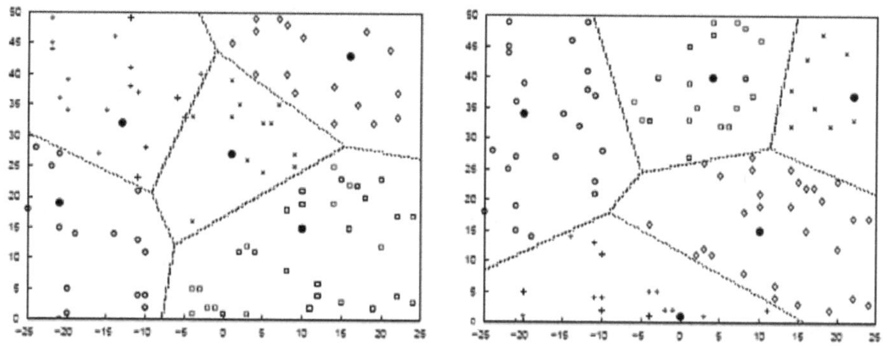

Fig. 19.5 Dynamic clustering in LEACH

In this paper, we review the parameters affected and algorithm based on wireless sensor networks. In many applications such as TSP [11], Vehicular Networks [14], Quadratic assignment problems [15], routing optimization in Telecommunication Networks, job shop scheduling (JSP), graph coloring problem (GRP) [16] solved by swarm intelligence based algorithms specially ACO and BCO. The research on ACO based and BCO based on routing protocols from last many years to found the optimal solution on different networking problems. The list of algorithms inspired from biologically based concept.

19.4.5 Classical Routing Protocol Approach

The classical routing protocol discuss with swarm-based routing protocol by Akkaya and Younis. In which the routing protocol such as LEACH and Directed diffusion approach are considered for the scalability and self-organization issues. LEACH (Low-energy adaptive clustering Hierarchy) is an efficient algorithm for self-organization which has the same concept of clustering approach. After forming clustering, each node will send the data to cluster heads and the cluster head schedules the node with TDMA and forwards it to the Base Station by direct communication. In directed diffusion, the sink broadcasts the interest message to each neighbor node with low data rate of gradient specifies the required data rates and direction given to the interested node. It is efficient for fault tolerance process in WSNs. The Figs. 19.5 and 19.6 show the dynamic clustering in LEACH protocol and the directed diffusion.

In MAC layer, the process of communication while forming the number of cluster for more number of sensor nodes in wireless senor networks and the communication between the intercluster and intracluster formation. In intercluster, the FDMA communication mode and in intracluster TDMA communication mode works. All the sensor nodes analogous to Ants and Bees search for next hops

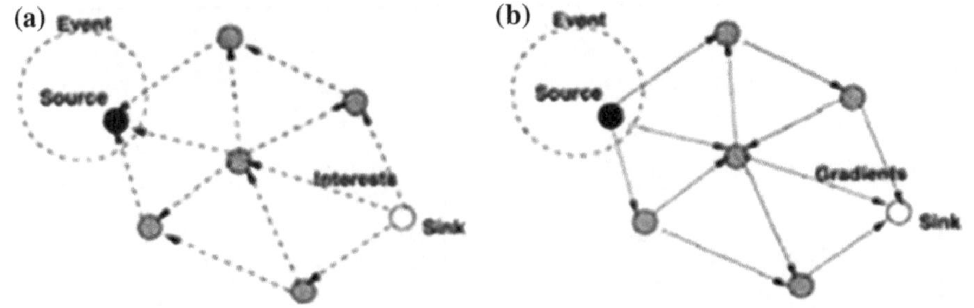

Fig. 19.6 A process for directed diffusion

communication. Those nodes that are nearer to the sources nodes with less radii form the number of groups as a cluster and then the cluster heads are selected by election algorithm for handling the data aggregation process (Tables 19.1 and 19.2).

19.5 Proposed Work

19.5.1 Biologically Inspired Algorithms

The biologically inspired methods such as ACO and BCO algorithms in which we consider three bioinspired algorithms for our study, i.e., BeeSensor which is the combination of BeeHive and Bee Ad hoc routing algorithms proposed from BCO and AntHOCnet as a AODV routing protocol which is designed especially for Mobile Wireless Ad hoc Sensor Networks, and self-organizing data gathering scheme (SDG). We try to design the novel architecture for self-organization based on the ability to adapt to topological changes in the network (changes such as join and leave operation in sensor networks). These algorithms are taken into consideration for comparison with the new proposed routing algorithms.

19.5.1.1 BeeSensor

Saleem and Farooq [10] have proposed Bee Sensor, a bee-inspired, reactive and event-driven multipath routing protocol for WSNs. Bee Sensor aims at energy efficiency, scalability, and long network lifetime. Energy efficiency is achieved by limiting the number of control messages, as well as of data packets through in-network aggregation. Paths are prioritized on the basis of their remaining energy levels to extend the network lifetime. In addition to forward and backward scout agents, Bee Sensor makes use of additional agents such as packers, foragers, and swarms. Packers receive data packets from the upper layers of the node architecture, and hand them over to a forager for transportation to a sink node.

Table 19.1 List of algorithms inspired by biological methods of ant colony optimization

Year	Authors	Algorithms	Problems studied
1991	Dorigo, Maniezzo and Colorni	AS	Traveling salesman
1995	Gambardella and Dorigo	Ant-Q	
1996	Dorigo and Gambardella	ACS and ACS-3-opt	
1997	Stutzle and Hoos	MMAS	
1997	Bullnheimwer, Hartl and Struss [14]	As_{rank}	
2000	Cordon et al.	BWAS	
1994	Maniezzo, colorni and Dorigo [15]	AS-QAP	Quadratic assignment
1997	Gambardella, Taillard and Dorigo HAS-QAP		
1997	Stutzle and Hoos	MMAS	
1998	Maniezzo	ANTS-QAP	
1999	Maniezzo and Colorni	AS-QAP	
1994	Colorni, Dorigo and Maniezzo	AS-JSP	Scheduling problem
1997	Stutzle	AS-FSP	
1999	Bauer et al.	ACS-SMTTP	
1999	Den Besten, Stutzle and Dorigo	ACS-SMTWTP	
2000	Merkle, Middendori and Schmeck	ACO-RCPS	
1997	Bullnheimwer, Hartl and Strauss [14]	AS-VRP	Vehicle routing
1999	Gambardella, Taillard and Agazzi	HAS-VRP	
1996	Schoonderwoed et al.	ABC	Connection-oriented network routing
1998	White, Pagurek and Opacher	ASGA	
1998	Di Caro and Dorigo	ANTNET-FS	
1998	Bonabeau et al. [7]	ABC-smart ants	
1997	Di Caro and Dorigo	AntNet and AntNet-FA	Connectionless network routing
1997	Subramaniam, Druchel and Chen	Regular ants	
1998	Heusse et al.	CAF	
1998	Vander put and Rothkrantz	ABC-backward	

(continued)

Table 19.1 (continued)

Year	Authors	Algorithms	Problems studied
1997	Gambardella and Dorigo	HAS-SOP	Sequential ordering
1997	Cost and Hertz	ANTCOL	Graph coloring
1998	Michel and Middendorf	AS-SCS	Shortest common supersequence
1998	Maniezzo and Carbonaro	ANTS-FAP	Frequency assignment
1998	Ramalhinho Lourenco and Serra	MMAS-GAP	Generalized assignment
1999	Leguizamon and Michalewicz	AS_MKP	Multiple knapsack
1999	Navarro Varela and Sinclair	ACO-VWP	Optical networks routing
1999	Liang and Smith	ACO-RAP	Redundancy allocation
2000	Solnon	Ant-P-solver	Constraint satisfaction

Table 19.2 List of algorithms inspired by biological methods of bee colony optimization

Year	Authors	Algorithms	Problem studied
1996	Yonezawa and Kikuchi	Ecological algorithms	Description of the collective intelligence based on bees behavior
1997	Sato and Hagiwara	Bee system (BS)	Genetic algorithm improvement
2001	Lucic and Teodorivic	BCO	Traveling salesman an problem
2001	Abbas	MBO	Propositional satisfiability problem
2002	Lucic and Teodorivic	BCO	Traveling salesman problem
2003	Lucic and Teodorivic	BCO	Vehicle routing problem in the case of uncertain demand
2003	Lucic and Teodorivic	BCO	Traveling salesman problem
2004	Wedde, Farooq, and Zhang	BeeHive	Routing protocols
2005	Teodorivic and Dell' Orco	BCO	Ride-matching problem
2005	Karaboga	ABC	Numerical optimizations
2005	Drias, Sadeg and Yah	BSO	Maximum weighted satisfiability problem
2005	Yang	Virtual bee algorithm (VBA)	Function optimizations with the applications in engineering problems
2005	Bentatchba, admane and Koudil	MBO	Max-sat problem
2006	Teodorovic, Lucic, Markovic and Delf Orco	BCO	Traveling salesman problem and a routing problems in networks
2006	Chong, Low, Sivakumar and Gay	Honey Bee colony algorithms	Job shop scheduling problem
2006	Plam, Saroka, Ghanbarzadeh, and Koc		Optimization of neural networks for wood defect detection
2006	Basturk and Karaboga	ABC	Numeric function optimization
2006	Navrat	BeeHive model	Web search
2006	Wedde, Timm and Farroq	BeeHiveAIS	Routing protocols
2007	Yang, Chen, and Tu	MBO	Improvement of the MBO algorithm
2007	Koudil, Benatchba, Tarabetand and El Batoul	MBO	Partitioning and scheduling problem
2007	Quijano and Passino	Honey Bee social algo	Solving optimal resources allocation problems
2007	Markovic, Teodorovic and Acimovic	BCO	Routing and wavelength assignment in all-optical networks
2007	Wedde et al.	BeeHive	Highway traffic congestion mitigation

(continued)

Table 19.2 (continued)

Year	Authors	Algorithms	Problem studied
2007	Karaboga and Basturk	ABC	Testing ABC algorithm on a set of multidimensional optimization problems
2007	Karaboga, Akay and Ozturk	ABC	Feed-forward neural networks training
2007	Afshar, Bozorg Haddada, Marin, Adams	Honey bee mating optimization	Single reservoir operation optimization problems
2007	Baykasoglu, Ozbakyr and Tapkan	Artificial Bee colony	Generalized assignment problem
2007	Teodorovic and Selmic	BCO	P-median problem
2008	Karaboga and Basturk	ABC	Comparison performances of ABC algorithm with the performance of other population based techniques
2008	Fathian, Amiri and Maroosi	Honey bee mating optimization algorithm	Cluster analysis
2008	Teodorovic	BCO	Comparison performance of BCO algorithm with the performance of other SI based techniques
2009	Pham, Haj Darwish, Eldukhr	Bees algorithm	Tuning the parameters of a fuzzy logic controller
2009	Davidovic selmic and eodorovic	BCO	Static scheduling of independent tasks on homogeneous multiprocessor system
2009–2013	Karaboga and Akay, Mala, Krishnanand, Li, Chu, Ruiz-vaneye, Wu, Rajeshar, Kashan, Mohammad and EI-ABD, Gao and Liu, Zou, Sunder Singh	ABC and modified ABC	Utilized for various applications and modified the ABC algorithms specially

In term, swarms transport a group of foragers back from the sink to the source node. Foragers are the main agents that transport events from the source to a sink node. Forward scouts carry the data, and are therefore launched on reactive basis. Intermediate nodes at Hl hops (or less) away from the source and deterministically broadcast them [17]. Here, we consider the BeeSensor from BCO for our research work.

19.5.1.2 Self-Organizing Data Gathering Scheme (SDG)

SDG protocol [18] aims to achieve scalability and reliability in sensor networks. In the protocol, a node uses another sink in case of failure. The protocol queries the fact that with a single sink, the sensor networks cannot tolerate energy, the sink remains isolated and the sensor network becomes useless as packets can no longer be routed to the sink. In the protocol, in order to minimize the routing overhead-agents are only generated by sink nodes in the form of backward ants, which are broadcasted by sink nodes on proactive basis. Sensor nodes communicate data and event information to their sink through the usual ACO techniques of stochastic forwarding. Node clustering in the algorithm is inspired from eggs and larvae grouping behaviors observed to their degree of similarity. Nodes at the borders of their cluster can dynamically change cluster membership according to a probabilistic mechanism that favors clusters with higher cluster pheromone [2].

19.5.1.3 AntHocNet

AntHocNet is based on the same principles thoroughly adapted to the challenges of MANETs. Nodes obtain routing information using ant-like agents which repeatedly sample and reinforce good paths. Routing information is maintained in arrays of pheromone variables, called pheromone tables. AntHocNet contains both reactive and proactive components: it is reactive in the sense that nodes only gather routing information for destinations which they are currently communicating with while it is proactive because nodes try to maintain and improve routing information as long as the communication is going on. Such hybrid architecture tries to combine the best of two worlds; while reactive algorithms are usually more efficient in terms of control overhead, proactive behavior allows better adaptively. Here is all for MANETS.

The design approach is based on following the horizontal categorization of self-organization (Fig. 19.7)

19.5.2 The Novel Architecture to be Proposed

See Fig. 19.8.

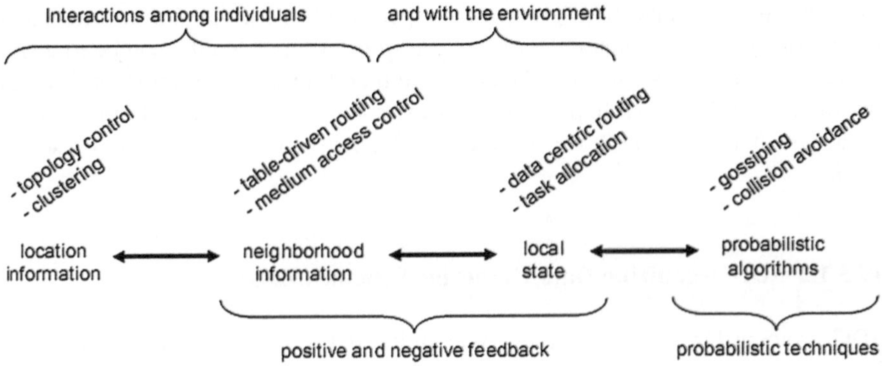

Fig. 19.7 Horizontal categorization of self-organization mechanisms in ad hoc sensor networks

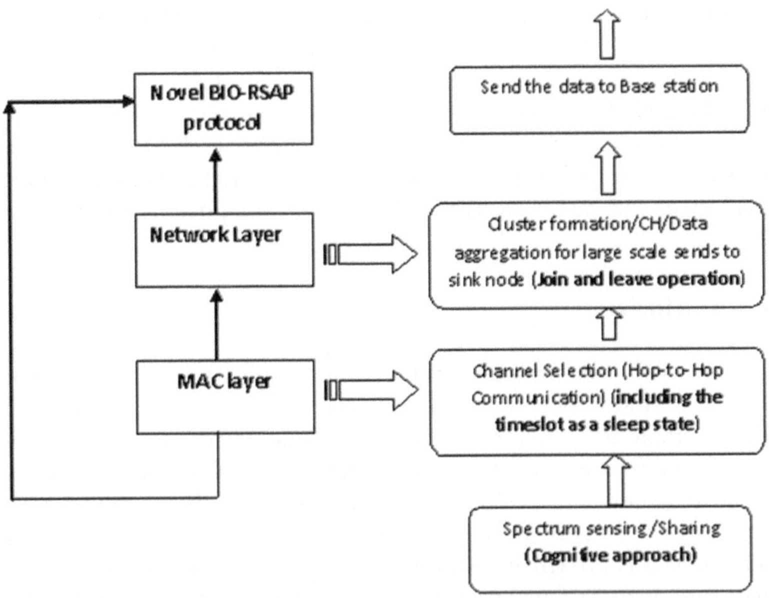

Fig. 19.8 Design approach of self-organized routing protocol

19.6 Conclusion

In this paper, we take a step to overview the different proposed issues and challenges of routing protocols in wireless sensor networks. We discuss the issues based on robustness in self-organization for better performance of routing in wireless sensor networks and provide the novel design approach of our research work using bioinspired methods.

Acknowledgments We thanks to all referenced authors for their research contribution as guidelines and valuable support for doing the research work and my guide Dr. L.G. Malik for guidance in research work.

References

1. Kiri, K., et al.: Robustness in sensor networks: difference between self-organized control and centralized control. Int. J. Adv. Networks Serv. **1**(2) (2009)
2. Zungeru, A.M.: Classical and swarm intelligence based routing protocols for wireless sensor networks: a survey and comparison. J. Netw. Comput. Appl. **35**, 1508–1536 (2012). (Elsevier)
3. Akyildiz, I.F., Kasimoglu, I.H.: Wireless sensor and actor networks: Research challenges. Elsevier Ad Hoc Netw. J. **2**, 351–367 (2004)
4. Karaki, J.N., Kamal, A.E.: Routing techniques in wireless sensor networks: a survey. Wireless Commun. IEEE **11**(6), 6–28 (2004)
5. Akkaya, K., Younis, M.: A survey of routing protocols in wireless sensor networks. Elsevier Ad Hoc Netw. J. **3**(3), 325–349 (2005)
6. Caro, G.D., Dorigo, M.: Antnet: Distributed stigmergetic control for communications networks, J. Artificial Intell. Res. **9**, 317–365 (1998)
7. Bonabeau, E., Dorigo, M., Theraulaz, G.: Swarm Intelligence: From Natural to Artificial Systems. Oxford University Press, New York (1999)
8. Li, K., Torres, C.E., Thomas, K., Rossi, L.F., Shen C.-C.: Slime mold inspired routing protocols for wireless sensor networks. Swarm Intell. **5**(3–4), 183–223 (2011)
9. Liu, M., Xu, S., Sun, S.: An agent-assisted QoS-based routing algorithm for wireless sensor networks. J. Netw. Comput. Appl. **35**(1), 29–36 (2012)
10. Saleem, M., et al.: BeeSesnor: a bee-inspired power aware routing protocol for wireless sensor networks. In: Proceeding of the 4th EvoCOMNET Workshop. LCNS, vol. 4448 (2007)
11. Darigo, M.: Optimization, learning and natural algorithms. Ph.D. dissertation (1992)
12. Saleem, K., et al.: Ant based self-organized routing protocol for wireless sensor networks. IJCNIS **1**(2) (2009)
13. Lucic, P., Teodorovic, D.: Bee system: modeling combinatorial optimization transportation engineering problems by swarm intelligence. In: Preprints of the TRISTAN IV Triennial Symposium on Transportation Analysis (pp. 441–445). Sao Miguel, Azones Islands, Portugal (2001)
14. Bullnheimwer, B., et al.: Applying the ant system to the vehicle routing problem. In: Paper presented at 2nd international conference on Metahauristics, Sophia Antipolis, France (1997)
15. Maniezzo, V., et al.: The ant system applied to the quadratic assignment problem. Technical Report IRIDIA/94-28, University Libre de Bruxelles (1994)
16. Costa, D., et al.: Ants can color graphs. J. Oper. Res. Soc. **48**(3), 295–305 (1997)
17. Chandni, et al.: Comparative analysis of routing protocols in wireless sensor networks. IJCSITS, ISSN: 2249-9555, **3**(1) (2013)
18. Kiri, Y., et al.: Self-organized data-gathering scheme for multi-sink sensor networks inspired by swarm intelligence (2007)
19. Zhang, Z., Long, K.: Self-organization paradigm and optimization approaches for cognitive radio technologies: a survey. IEEE Wireless Commun. **20**(2), 36–42 (2013)
20. Mills, K.L.: A brief survey of self-organization in wireless sensor networks. Wirel. Commun. Mob. Comput. **7**, 1–12 (2007)
21. Bitam, S., et al.: A survey on bee colony algorithms. IEEE (2010)
22. Capella, J.V., et al.: A new robust, energy-efficient and scalable wireless sensor networks architecture applied to a wireless fire detection system. IEEE Computer Society, pp. 395–398. (2009)
23. Yazdi, F., et al.: Ant colony with colored pheromone routing for multi objective quality of services in WSNs. Int. J. Res. Comput. Sci. ISSN 2249-8265 **3**(1), 1–9 (2013)
24. Hentefeux, F., et al.: Self-organization protocols behavior during the sensor networks life. (PIMRC'07), IEEE (2007)

25. Ali, Z., et al.: Critical analysis of swarm intelligence based routing protocols in Ad hoc and sensor wireless networks. IEEE, pp. 287–292. (2011)
26. Abbasi, A.A., et al.: A survey on clustering algorithms for wireless sensor networks. Comput. Commun. **30**(14), 2826–2841 (2007)
27. Younis, O., et al.: Distributed clustering in Ad hoc sensor networks: a hybrid, energy efficient approach
28. Atakan, B., et al.: Biologically-inspired spectrum sharing in cognitive radio networks. IEEE (2007)
29. Li, G., et al.: Enhanced biologically inspired spectrum sharing for cognitive radio networks. IEEE (2010)
30. de Doenico, A., et al.: A survey on MAC strategies for cognitive radio networks. IEEE (2010)
31. Teodorovic, D., et al.: Bee colony optimization: the application survey. ACM Trans. Comput. Logic **1529**, 1–20 (2011)
32. Aksa, K., et al.: A comparison between geometric and BIO-Inspired algorithm for solving routing problem in WSN. IJNC **2**(3), 27–32 (2012)
33. Wede, H.F., et al.: BeeHive: An efficient fault tolerance routing algorithm inspired by honey bee behavior. Springer, Heidelberg (2004)
34. Zheng, C., et al.: A survey on biologically inspired algorithms for compute networks. IEEE (2013)
35. Di Caro, G.A., et al.: Bio-inspired techniques for self-organization in dynamic networks (BISON) (2005)
36. Dressler, F.: A study of self-organization in ad hoc and sensor networks. Comput. Commun. **31**, 3018–3029 (2008). Elsevier
37. Hong, J., et al.: Towards bio-inspired self-organization in sensor networks: applying ant colony algorithm. 1550-445X, IEEE (2008)
38. Paone, M., et al.: A multi-sink swarm based routing protocol for WSN. 978-1-4244-1, IEEE (2009)
39. Paone, M., et al.: A Swarm-based routing protocol for wireless sensor networks. IEEE (2007)

Chapter 20
Energy Efficient Fuzzy Clustering in Wireless Sensor Network

Suman Bhowmik and Chandan Giri

Abstract The nodes used in wireless sensor network (WSN) are mostly battery operated and so energy is the most precious resource in any WSN application. Therefore, we need to develop algorithms which are energy efficient. In most WSN applications node density is made very high. Energy can be conserved if it is possible to turn off a substantial number of redundant nodes. Using clustering technique it is possible to turn off those nodes. In this chapter, we propose a fuzzy based, energy efficient clustering algorithm. We also find the time and communication complexities of the algorithm.

Keywords WSN • Fuzzy clustering • Clustering algorithms

20.1 Introduction

The tiny sensor nodes used in wireless sensor network (WSN) are mostly battery operated and sustained operation will drain out the energy of nodes and reduce the lifetime of the network. Sensor nodes are very cheap, so in most applications a very high-density deployment is used. Communication module of sensor node consumes most energy. So, turning off that part can conserve a significant amount of energy. Partitioning the whole deployment area into some clusters can achieve the goal. Cluster head (a special node) can be used as a gateway of a cluster so

S. Bhowmik (✉)
College of Engineering and Management, Kolaghat, West Bengal, India
e-mail: bhowmik.suman@gmail.com

C. Giri
Bengal Engineering and Science University, Shibpur, West Bengal, India
e-mail: chandangiri@gmail.com

R. Maringanti et al. (eds.), *Proceedings of Ninth International Conference on Wireless Communication and Sensor Networks*, Lecture Notes in Electrical Engineering 299,
DOI: 10.1007/978-81-322-1823-4_20, © Springer India 2014

that all the cluster members can send their collected data to the cluster head and the cluster head in turn will forward those data to the sink node. So for most of the time, we can switch off the transceiver part of the cluster member nodes to conserve energy.

Related Work: In recent years, a large number of clustering algorithms are proposed for WSNs. The algorithm proposed in [1] elects cluster heads solely based on probability. The single hop communication model is used by the cluster heads to forward packets to the base station. Authors in [2] proposed another version of unequal LEACH (ULEACH), which uses energy ratio (current energy to initial energy) and competition radius to elect cluster heads. Authors in [2] proposed an algorithm to solve the hotspot problem by creating unequal sized clusters varying the competition radius. Small-sized clusters are formed near to the base station and as we go away from it larger clusters are created. In their paper, [3] proposed an energy-driven unequal clustering (EDUC) algorithm which rotates of the role of CH based on either time or energy. Authors in [4] introduced an unequal variant of HEED (UHEED). Comparison of UHEED shows that for most of the case it performs better than the HEED [5], LEACH [1], and ULEACH [2] algorithms.

The chapter is organized as follows: this section starts with the introduction and related works on clustering so far. Section 20.2 explains the proposed clustering algorithm. Section 20.4 analyzes the complexities of the algorithm. Section 20.5 shows simulation results and compares performance of proposed algorithm with some existing well-known algorithms. Finally, Sect. 20.6 draws the conclusion.

20.2 Clustering Requirements

We have used the fuzzy communication model in [6]. Around a transmitting node five fuzzy regions are formed depending on five fuzzy distances of the receiver as “very near,” “near,” “moderate,” “far,” and “very far.” The constraints on algorithm are (1) It must be a fully distributed one. (2) No synchronization between nodes is required. (3) Base station will be in one hop distance from at least one cluster head. (4) Prefer higher ranked nodes as cluster head and connector. (5) All cluster members must be within one hop distance from cluster head. (6) Each node will belong to only one cluster except connector node. (7) Cluster members are selected within far distance from cluster head. (8) Cluster heads will be out of communication range with each other. (9) Neighboring clusters communicate via connector nodes. (10) A connector node will connect not more than two clusters. (11) Two neighboring clusters will be connected via only one connector node.

With these constraints, the whole deployment region will be partitioned into a number of overlapped clusters. After clustering, active nodes will form a connected graph as shown in Fig. 20.1. These requirements identify the following categories of nodes (see Fig. 20.2):

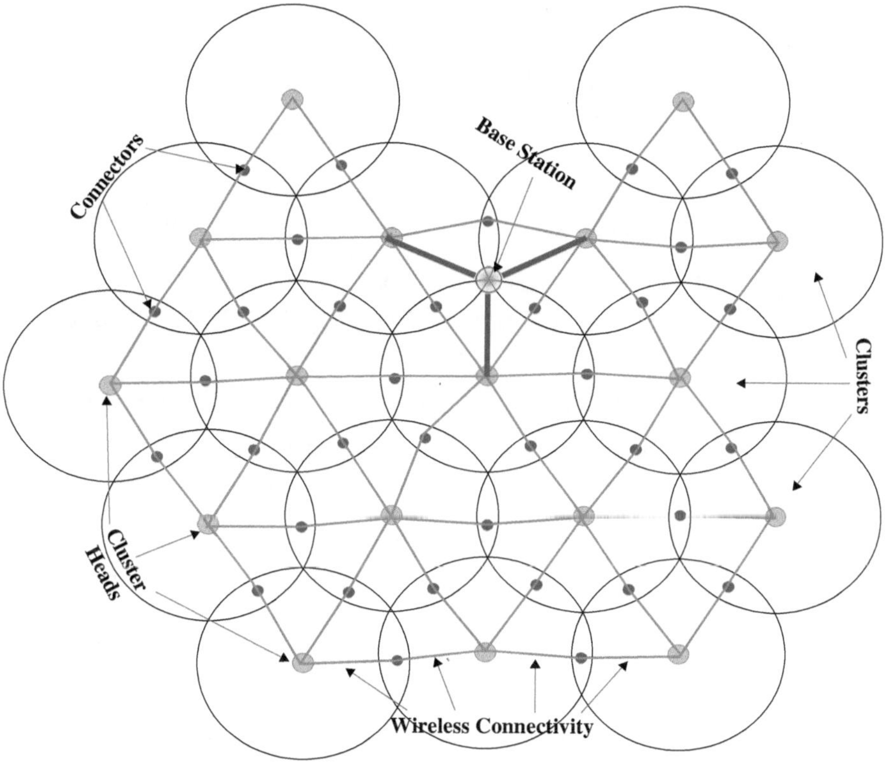

Fig. 20.1 Graph formed after clustering

Base Station: Central computer with large resources, where collected data is sent for further processing. **Cluster Head (CH)**: Remains active throughout a single round of iteration carry forward data collected from member nodes toward the base station. **Member Node**: Any node other than the cluster head is treated as a member node. The member nodes can be further categorized as: **Permanent Member (PM)**: Nodes received either a single *join* message from a CH or being not selected as connector node after exchanging *select* message. These nodes remain in sleep mode throughout one round. **Tentative Member (TM)**: Nodes received more than one *join* message and not yet become permanent member or connector. These nodes exchange *select* message to determine whether those will be a permanent member of a cluster or connector between two clusters. **Connector Node**: TMs which receive more than one *join* message and their rank is highest among the other TMs involved in contention when exchanging *select* message. These nodes connect two adjacent clusters and move any message between the two clusters.

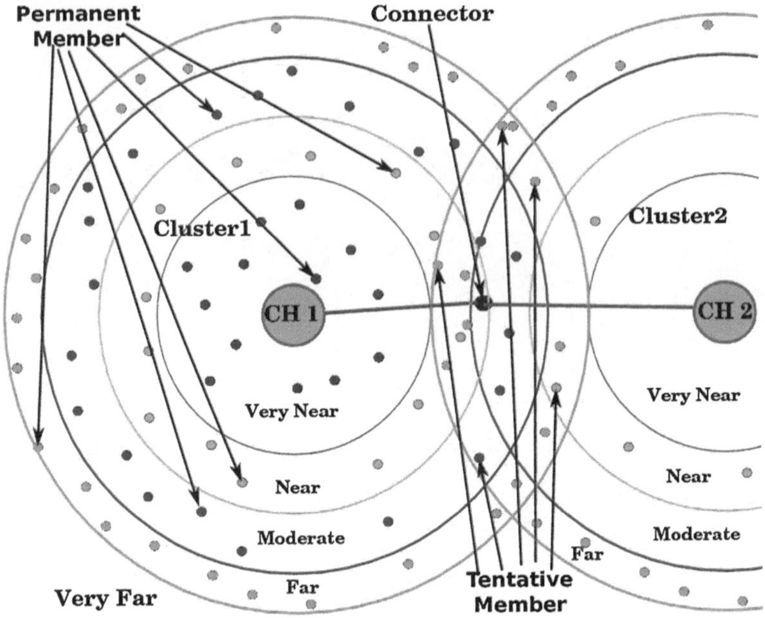

Fig. 20.2 Sensor node categories

Table 20.1 Linguistic values of input and output variables

Input		Output
remain Energy	timesUnUsed	Rank
Empty	Immediately used	Highest
Small	Recently used	Very high
Moderate	Long used	High
Large	Very long used	Medium
Very large	Unused	Moderately low
Full		Low
		Extremely low
		Nil

20.3 Algorithm

Each node executes the algorithm in an indefinite loop throughout its lifetime. Function Main() shows one such iteration. First a node finds its own fuzzy rank. Then find neighbors using *hello* message until it becomes a non-normal node. From H number of neighbors, a node elects itself as CH (calls clusterHead()) either if its rank is highest or if its rank is equal to maximum rank found and then if its *nodeID* is minimum. Otherwise, wait for Δt_2 time to receive J number of *join* message from already elected cluster heads. If a normal node receives at least one *join* message, it sets itself as TM (calls tentativeMember()) (Table 20.1).

Algorithm 1: Distributed clustering Functions

```
Function Main(Node n_i){
  calculateRank()
  while n_i.nodeType == NORMAL do
    findNeighbors(n_i)
    n_j = node with maximum rank among H neighbours
    if (n_i.rank > n_j.rank) or ((n_i.rank == n_j.rank) and
   (n_i.nodeID < n_j.nodeID)) then
      n_i.nodeType = CH /* cluster head */
    else
      wait for Δt_2 time
      receive J number of join message
      if j(∈ J) < 0 then
        n_i.nodeType = TM /* Tentative member node */
  if n_i.nodeType = CH then
    clusterHead(n_i)
  if n_i.nodeType = TM then
    tentativeMember(n_i, J join message)
}
Function findNeighbors(Node n_i){
  createHelloMessage(n_i.rank, n_i.niodeID)
  broadcast hello message with specified power
  wait Δt_1 time
  receive H number of hello message
  for ∀h ∈ H do
    fd_h = findFuzzyDistance(h)
    store hello message h with fd_h in neighbor[i]
}
Function calculateRank(Node n_i){
  remainEnergy = findFuzzyEnergy(n_i.batteryRemainEnergy)
  timesUnUsed = findFuzzyTimesUnused(n_i.timesNotUsed)
  findFuzzyRank(remainEnergy, timesUnUsed)
}
Function clusterHead(Node n_i){
  createJoinMessage(n_i.nodeID)
  broadcast join message with specified power
  wait for Δt_3 time
  receive N number of notify message
  for ∀n ∈ N do
    neighborCluster[i].clusterID = n.clusterID
    neighborCluster[i].connectorID = n.connectorID
  n_i.state = ACTIVE
  n_i.timesNotUsed = 0
}
```

Algorithm 2: Remaining functions

```
Function permanentMember(Node n_i, join message j){
   parent.clusterID = j.senderID
   n_i.fuzzyDistance = findFuzzyDistance()
   reduce transmission power according to fuzzy distance to cluster head
   n_i.nodeType = PM
   n_i.timesNotUsed++
   n_i.state = SLEEP
}
Function tentativeMember(Node n_i, J join message){
   if J == 1 then
   |    permanentMember(n_i, join message)
else
   |    for ∀j ∈ J do
   |    |    fd_j = findFuzzyDistance(j)
   |    j_min = join message with minimum fuzzy distance
   |    j_nmin = join message with next minimum fuzzy distance
   |    createSelectMessage(j_min, j_nmin, n_i.rank)
   |    broadcast select message with double power
   |    wait for Δt_4 time
   |    receive S number of select message
   |    find s select message that matches the sent message (s ∈ S)
   |    s_m = select message with maximum rank from s message
   |    if (n_i.rank > s_m.rank) or ((n_i.rank == s_m.rank) and
   |  (n_i.nodeID < s_m.senderID)) then
   |  |    connectorNode(n_i, select message s broadcasted)
   |  else
   |  |    permanentMember(n_i, j_min join message)
}
Function connectorNode(Node n_i, select message s){
   createNotifyMessage(s.cluster1, s.cluster2)
   send notify message to both s.cluster1 and s.cluster2
   neighborCluster[0].clusterID = s.cluster1
   neighborCluster[1].clusterID = s.cluster2
   fd_min =find minimum fuzzy distance from n_i to s.cluster1 and s.cluster2
   reduce transmission power according to fd_min
   n_i.nodeType = CONNECTOR
   n_i.timesNotUsed = 0
   n_i.state = ACTIVE
}
```

Rank (calculateRank()) of a node depends on remaining battery energy (*remainEnergy*) and the time the node remains unused (*timesUnUsed*). At the beginning of each round, a node finds linguistic values of *remainEnergy* and *timesUnUsed* (see Table 20.2) and calculates the *rank* using the inference rules (see Table 20.2). To **find Neighbors** (findNeighbors()) after calculating the *fuzzy rank*, each node broadcasts a *hello* message with its *rank* and *nodeID*. Within Δt_1 time each node receives H number of *hello* message form neighbors. For each

Table 20.2 Inference rules to find the rank of a node

Input		Output
remainEnergy	timesUnUsed	Rank
Empty	Any	Nil
Small	Immediately used, recently used	Extremely low
Small	Long used, very long used	Low
Moderate	Immediately used, recently used	
Small	Unused	Moderately low
Moderate	Long used, very long used	
Large	Immediately used	
Moderate	Unused	Medium
Large	Recently used, long used	
Very large, full	Immediately used	
Large	Very long used, unused	High
Very large, full	Recently used, long used	
Very large, full	Very long used, unused	Very high

message received, calculate the fuzzy distance from the sender using technique discussed in [6]. **A CH** (clusterHead() called from Main()) first broadcasts a *join* message and waits for Δt_3 time to receive *N* number of *notify* messages from *N* number of connector nodes. Store neighbor CH information in neighborCluster array from *notify* messages. Then the cluster head resets the variable *timesNotUsed* to identify that recently it is being used as an active node. **A TM** (tentativeMember() called from Main()) receives *J* number of *join* messages from *J* number of newly elected CH. If the node receives only one message it becomes a PM of that cluster head. If it receives more than one message, find two message for which the fuzzy distance of the node are minimum and next to minimum from corresponding CH. The node then broadcasts a *select* message (with its *rank* and *id* of this two CH) with double transmit power. After waiting for Δt_4 amount of time, a TM receives *S* number of *select* messages from other TMs. Elects itself as a connector if its rank is maximum among the *select* messages which matches with its own. In case of tie, minimum id node is elected. Otherwise, it will make itself as a permanent member to the cluster whose fuzzy distance is minimum. **A PM** (permanentMember()called from tentativeMember()) first sets the sender of the *join* message as its parent, saves fuzzy distance from the cluster head, and reduces the transmission power according to the fuzzy distance. Then it increments *timesNotUsed* to increase the possibility of being selected as either a cluster head or connector in the next iteration and go to sleep mode. Being elected as a **connector node** between two cluster heads specified in matching *select* message, it sends a *notify* message (with both CH information and its own *id*) to the two cluster heads. Save the two cluster information to *neighborCluster* array, calculate the minimum fuzzy distance ($fd_{\min}$) between both the cluster heads, and reduce its transmission power according to $fd_{\min}$. Reset the variable *timesNotUsed* to identify that recently it is being used as an active node.

20.4 Complexity Analysis

Lemma 1 *On average, number of neighbours of a node within its transmission range of radius r deployed randomly in a X × Y field with uniform probability distribution is proportional to total number n of nodes deployed.*

Proof The probability of falling a node within the rectangular field is $\frac{1}{XY}$. For a particular node, the neighbors are those nodes which fall within its transmission range of radius r centered at the node. Probability of falling nodes within this circular region is $\frac{\pi r^2}{XY}$. So for n nodes, the expected number of nodes belonging to this region becomes $\frac{\pi r^2}{XY} n$. Thus, on average, the number of neighbors of a node is proportional to the total number of nodes deployed. □

Lemma 2 *The time complexity of the clustering algorithm for n number of sensor nodes deployed is O(n).*

Proof From algorithm we see that rank calculation can be done in $O(1)$ time. findNeighbors() processes H number of hello messages so its contribution is $O(H)$. clusterHead() process N number of *notify* messages in a single loop, thus its contribution is $O(N)$. permanentMember() and connectorNode() both contribute $O(1)$. tentativeMember() first processes J number of *join* messages in a single loop and then processes S number of *select* messages in another loop so its contribution is $O(J) + O(S)$. Now Main() processes J *join* messages (if not a cluster head) in a loop to contribute $O(J)$. So, time complexity becomes $O(H) + O(N) + 2 \times O(J) + O(S)$. Since $J < N \ll S \ll H$, the overall time complexity becomes $O(H)$. From lemma 4 we see that $H \propto n$; so the computational complexity becomes $O(n)$. □

Lemma 3 *Communication or message complexity of clustering algorithm for n number of sensor nodes is O(n).*

Proof The algorithm uses four type of messages: (1) *hello*, sent once by each node in the network, generates $O(n)$ messages. (2) Each elected CH broadcasts *join* message once. Thus, m number of CH formed in each round, generate $O(m)$ messages. (3) Each tentative member (TM) broadcasts one *select* message, thus p number of TM contribute $O(p)$ message. (4) Each connector sends two *notify* messages, thus q number of connectors contribute $O(q)$ message. So, overall complexity becomes $O(n) + O(m) + O(p) + O(q)$. Number of cluster heads and connector nodes elected in a round is much less than the total number of nodes. *Select* message is used to elect connector node among the tentative member around two clusters, i.e., it is a local phenomena and confined in a small region. Therefore, we can write $m, q \ll p \ll n$. So, the overall communication complexity is $\approx O(n)$. □

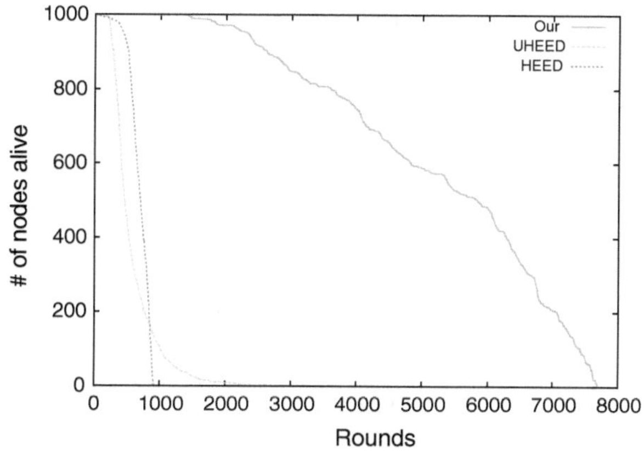

Fig. 20.3 Lifetime comparison with 1,000 nodes in 500 × 500 area

20.5 Simulation Results

For simulation, we assume the same energy model as defined in [7]; these are $E_{\text{elec}} = 50$ nJ/bit, $\varepsilon_{fs} = 10$ pJ/bit/m^2, $\varepsilon_{mp} = 0.0013$ pJ/bit/m^4, and $E_{DA} = 5$ nJ/bit/signal. For all the cases discussed, we assume that the sink node is at (50, 175). A node is considered dead if its battery energy is depleted to 99.9 %.

Case 1 In this case, we have done simulation with 1,000 nodes deployed in a 500 × 500 m field, initial battery energy used 2 J, data packet size was 1,000 bits, and cluster radius taken 35 m. The result of comparison with HEED [5] and UHEED [4] algorithms is shown in Fig. 20.3. In this deployment, the first node fails at 512, 24, and 40 rounds in our, UHEED, and HEED algorithms, respectively. The half nodes fail at 5,842, 462, 702 rounds in the three algorithms, respectively and the last node fails at 7,691, 3,321, and 912 rounds in the respective algorithms. This clearly shows that our algorithm gives far more better results than these two algorithms.

Case 2 We compare our algorithm with EECS [8], LEACH-E [7], and HEED [5] (presented data taken from [8]) on distribution of number of clusters generated in different rounds. We deployed 1,000 nodes in 200 × 200 m field, initial battery energy taken 1 J, data packet size was 4,000 bits, and cluster radius taken 40 m. We have taken randomly selected 100 rounds and found the distribution (see Fig. 20.4). The second and third figure in Fig. 20.4 shows the results for EECS and LEACH-E which depict that there are wide variations in the number of clusters generated whereas in case of HEED and Our algorithm (fourth and first figure in Fig. 20.4, respectively) there are very small variation. Moreover, compared to HEED, our algorithm behaves much better as the number of cluster heads generated is concentrated more on value (55 rounds) whereas for HEED it is almost evenly distributed on (36 rounds) and 7 (42 rounds). Thus, our algorithm gives more predictable behavior.

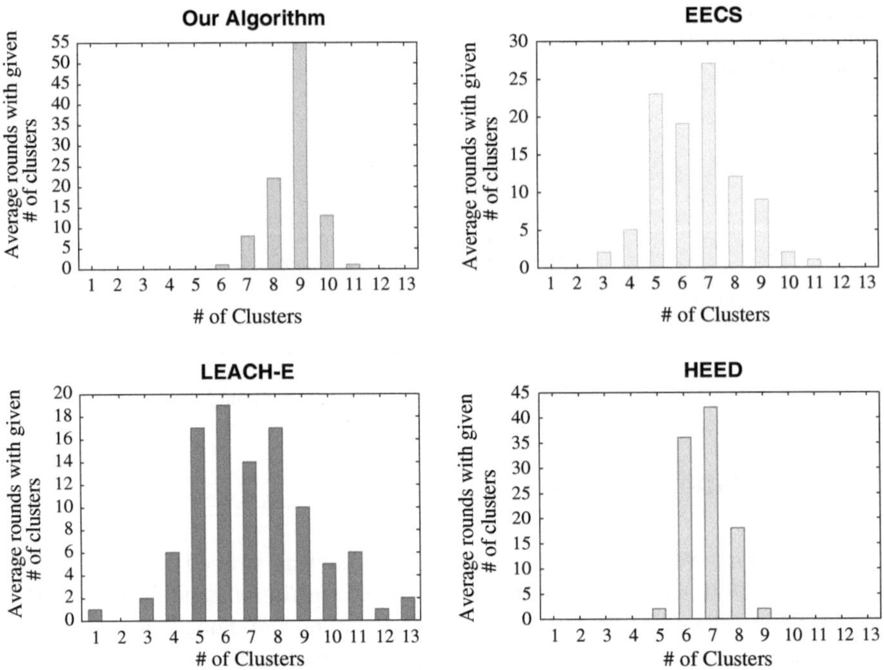

Fig. 20.4 Number of clusters in different algorithms (Our, EECS, LEACH-E, and HEED)

20.6 Conclusion

In this paper, we proposed one energy efficient clustering algorithm. To make the clustering energy efficient in the first hand, we used unequal sized clustering technique rather than the equal sized one; in the second hand, we have used both energy-driven, rank-based as well as time driven (in rounds) cluster head election techniques. The simulation of the algorithm shows that it gives far better results than the existing algorithms.

References

1. Heinzelman, W., Chandrakasan, A., Balakrishnan, H.: Energy-efficient communication protocol for wireless microsensor networks. In: Proceedings of the 33rd Annual Hawaii International Conference on System Sciences, vol. 2, p. 10 (2000)
2. Ren, P., Qian, J., Li, L., Zhao, Z., Li, X.: Unequal clustering scheme based leach for wireless sensor networks. In: Fourth International Conference on Genetic and Evolutionary Computing (ICGEC), pp. 90–93 (2010)
3. Yu, J., Qi, Y., Wang, G.: An energy-driven unequal clustering protocol for heterogeneous wireless sensor networks. J Control Theory Appl **9**(1), 133–139 (2011)

4. Ever, E., Luchmun, R., Mostarda, L., Navarra, A., Shah, P.: Uheed—an unequal clustering algorithm for wireless sensor networks. In: SENSORNETS, pp. 185–193 (2012)
5. Younis, O., Fahmy, S.: Heed: a hybrid, energy-efficient, distributed clustering approach for ad hoc sensor networks. IEEE Trans. Mob. Comput. **3**(4), 366–379 (2004)
6. Bhowmik, S., Giri, C.: Fuzzy communication model for sensors in wireless sensor network. In: International Conference on Communications, Devices and Intelligent Systems (CODIS), pp. 254–57, Dec 2012
7. Heinzelman, W., Chandrakasan, A., Balakrishnan, H.: An application-specific protocol architecture for wireless microsensor networks. IEEE Trans. Wireless Commun. **1**(4), 660–670 (2002)
8. Ye, M., Li, C., Chen, G., Wu, J.: EECS: an energy efficient clustering scheme in wireless sensor networks. In: 24th IEEE International Performance, Computing, and Communications Conference. IPCCC 2005. pp. 535–540 (2005)

Chapter 21
Network Optimization in WSN's

Rakhi Khedikar, Avichal Kapur and M. D. Chawan

Abstract Recently wireless sensor network (WSN) is very popular for monitoring the remote or hostile environments. One major challenge in WSN is to build energy efficient network with total coverage & connectivity. The WSN consist of a large number of tiny nodes having sensing, computation & communication capability. The deployment of sensor nodes is the first step in establishing a sensor network. The large number of sensor nodes are deployed randomly & must be in the form of cluster, as a location of each particular node cannot be fully guaranteed a priori so that nodes must completely cover the target area. To build energy efficient network with the stochastically placed nodes, it is necessary to activate only vital number of nodes at any particular moment. In this paper we introduce a heuristic that will find out the mutually exclusive set called disjoint complete set of sensor nodes, where the member of each set together will completely cover the target area and work by turns. As the more number of sensor covers can be found the sensorsnetwork lifetime can be prolonged.

Keywords: Wireless sensor networks • Connectivity • Coverage • Energy efficiency • Disjoint complete set

21.1 Introduction

Wireless sensor network (WSN) build by a tiny nodes having sensing computation & communication capabilities, each nodes well-found with center processing unit (CPU), battery, sensors and radio transceiver network through which nodes

R. Khedikar (✉)
Research Scholar, RTMNU, Nagpur, Maharashtra, India
e-mail: rkhedikar@gmail.com

A. Kapur
MGI, Nagpur, Maharashtra, India

M. D. Chawan
RCOEM, Nagpur, Maharashtra, India

R. Maringanti et al. (eds.), *Proceedings of Ninth International Conference on Wireless Communication and Sensor Networks*, Lecture Notes in Electrical Engineering 299,
DOI: 10.1007/978-81-322-1823-4_21, © Springer India 2014

established wireless connection, processing and transmission of data collected by the sensor nodes. The large number of sensors node deploy in a field to cover the target area for performing the task like animal tracking, environmental monitoring, military surveillance etc. WSN is most often set up in an ad hoc mode by means of small-size identical devices grouped into nodes distributed densely over a significant area. As the sensor nodes are battery operated and the size of cell is small so the amount available of energy of the node is also limited. Therefore, one of the important concerns in wireless sensor network is to save the energy by developing some new deployment strategies, architecture and application with low energy consumption. From the literature survey we observed that to obtain full coverage with minimal energy consumption sthere is the problem of placement of sensor nodes. The sensor node operates between two different operating modes i.e. active mode and sleep mode. When the node operates in active mode it performs full operations like sensing, computation and communication. To perform those operations sensors consume a lot of energy while sensor operates in sleep mode it consumes comparatively less energy. To save the energy of WSN we need to operate the sensor in active and sleep mode periodically. With consideration of the problem we have develop the algorithm that organized the number of sensors available in a network into mutually exclusive sets and each set of the network complete cover the monitor area.

At any movement only one such set is active and consumes power. In another specified interval another set is activated and first one is deactivated this process will repeated till whole sets are used and the process repeats until the sensors are out of power. From the above discussion it is observed that the network lifetime depends on the number of allocated sets; therefore, more number of sets increasing the lifetime of system hence the aim of the algorithm is to create maximum number of disjoint complete sets.

The sensor node resides some processing circuit and wireless and a wireless transceiver. In remote and inhospitable areas the large numbers of sensor nodes are deployed to cover the There are two types of deploying methods are used to cover the target area i.e. controlled deployment and random deployment [1]. In controlled deployment the sensors are deploy with well design plan and such type of deployment plan will applicable in small areas. The aim of such deployment methods use a limited number of sensors in an area to reduce the system cost whereas in random deployment the sensors are randomly deployed in a large area and such type of deployment are applicable in a large inaccessible areas [2–7]. Once the sensor network deployed the sensor node organize the wireless network and send the observations (necessary data) to the base station by combine the observations the sensor network provides global view of monitor area to user. We assume for simulation purpose the sensing area of a sensor is circle and the radius of the circle is equal to the sensing range of a sensor.

The rest of the papers organized like first we discuss the approaches to design the energy efficient network consume less energy. The short description of communication methods, energy conservation techniques (power save deployment methodology) and algorithms for computing the optimal transmitting ranges in order to generate a network with desired properties while reducing sensors energy consumption (Scheduling) is provided power save techniques attempt to save nodes energy by putting its radio transceiver in the sleep state. Finally, we discuss the idea of our novel location based power save scheme utilizing hierarchical structure with periodic coordination of network nodes activity.

21.2 Related Work

The scheduling is one of the effective methods to increase the lifetime of wireless sensornetwork [8, 9]. There are two operating modes of sensors active mode and sleep mode. When sensor operates in active mode performs full operations like sensing, computation and communication and to perform those operations sensor consumes enormous energy and when sensors are in sleep mode it consume relatively small energy. Hence by scheduling we operate the sensors into two different operating modes active mode and sleep mode with a proper intervals.

In the literature the various methods have been proposed on sensing coverage and connectivity for organizing energy efficient wireless sensor network [10–12]. To cover the complete target area two types of deploying methods are used i.e. controlled deployment and random deployment [1]. In control deployment the sensors are deploy with well-design plan and such type of deployment plan will be applicable in small areas. The aim of such type of deployment methods use of smallest number of sensors in a limited area to reduce the cost of the system [2–7, 9, 13, 14], whereas some methods consider dispatching a set of mobile sensors to satisfy the coverage and connectivity requirements [1, 2, 9, 13–20]. In some case when the target area and the number of sensor used to cover the target area is large hence sometimes it is very difficult to apply control deployment method to place the sensor positions prearrange in such cases random deployment method is used to place the sensors to the target area. In case of random deployment the number of sensors deployed densely to get the total coverage in large target areas.

By applying different scheduling rules we change the operating mode of the wireless sensor node. As the subset of the sensors cover the target area completely, the other sensors can be schedule to be in the sleep mode to save the energy. For example, in the randomized scheduling methods in [20], sensors are randomly assigned to multiple working subsets of sensors. For each subset of sensors, the algorithm used an extra-on rule for guaranteeing network connectivity and then updated the working schedule accordingly. Lin and Chen [16] later improved the approach of [20] by detecting and eliminating coverage holes in the subsets. Abrams et al. [21] designed three approximation algorithms for a variation of the SET k-cover problem. The work in [2–7, 14, 21–26] also provides methods for constructing working subsets of sensors, but their solutions trade off complete coverage in exchange for prolonging the network lifetime. Slijepcevic and Potkonjak [24] proposed a greedy deterministic approach called the "most constrained–minimally constraining covering (MCMCC)" heuristic to completely cover the target area. MCMCC cannot guarantee finding the optimum, but it works much faster than MC-MIP for problems in a large scale. In MCMCC, a function is defined favoring the sensor which covers the most constrained field, whereas the other fields covered by the sensor are minimally constraining. In [9] Cardie and Due proposed a "maximum covers using mixed integer programming (MC-MIP)" algorithm to find the maximum number of disjoint complete cover sets for covering a set of target points. They transformed the problem into a maximum flow problem and then formulated it as a mixed integer programming. By using a branch and bound method, MC-MIP acts as an implicit exhaustive search which guarantees finding the optimal solution. However, as the numbers of sensors and targets become larger, the running time of MC-MIP increases exponentially. Whether a field is constrained or not depends on the number of sensors that can cover the field. Each complete cover set in MCMCC is constructed by selecting sensors according to the heuristic objective function. MCMCC can be applied for point-coverage and area coverage but MC-MIP address only point-coverage problem. As area coverage involves a much larger number of coverage targets then point coverage and each field in the target area must be completely covered. The aim of proposed algorithm is to find out the maximum number of disjoint complete set for maximizing the lifetime of wireless sensor network. The proposed algorithmcan be applicable to both point-coverage and area-coverage disjoint set covers problems. Results show that the proposed algorithm can achieve high-quality solutions with a much faster optimization speed.

21.3 Problem Definition and Discussion

(a) Cover set

The lifetime of wireless sensor network increases by finding the maximum number of disjoint complete set. Suppose in L × W area a set S = {s1, s2… sN} of sensors are deployed. The objective of the sensor set covers problem is to find the maximum number T of disjoint complete cover sets and the corresponding cover sets Si, satisfying:

1. Each set Si = {si1, si2… si|Si} ⊆ S of sensors forms complete coverage to the target area, where |Siis the number of sensors that are activated in the ith schedule, i = 1, 2, … T;
2. Each sensor belongs to no more than one cover set that is Si

$$Si \cap Sj = \varphi \tag{21.1}$$

where $i = j, i, j = 1, 2, \ldots T$;

As each sensor Si monitors a certain area, complete coverage to a target area means that the whole area is under- monitored.

Figure 21.1 shows the complete coverage area covered by the sensors. Each sensor Si cover the area within the sensing range ri. Practically sensing range covered by the sensors is not circular shape but can be any irregular shape. In this

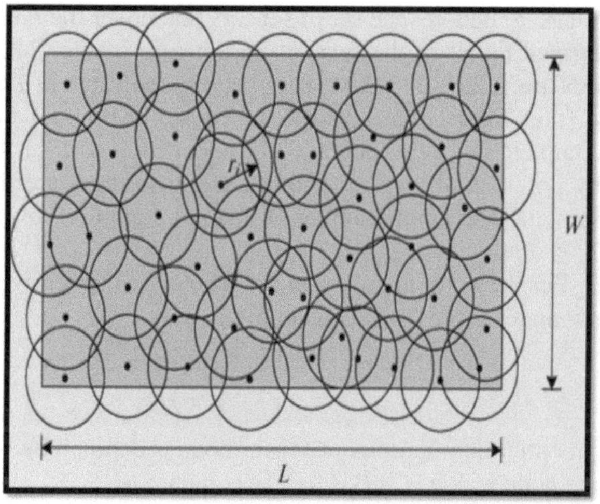

Fig. 21.1 Illustration of complete coverage to an L × W area. Each sensor *si* is marked as a *dot* and has a sensing range *ri*. Each *circle* represents the sensing area of the corresponding sensor

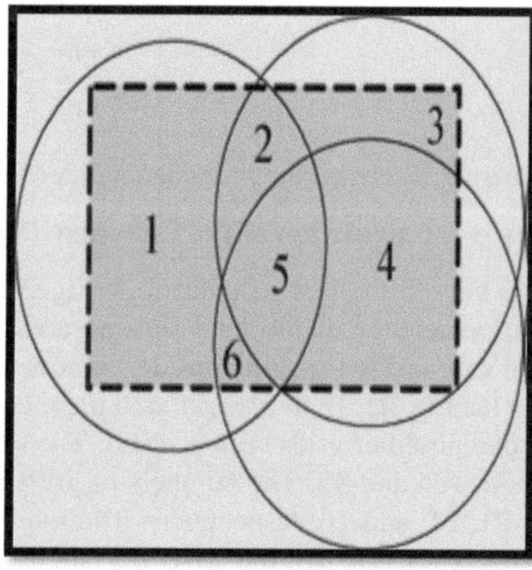

Fig. 21.2 Example of six fields formed by three sensors. The area contained in the same field is covered by the same set of sensors by sensors

paper we simply consider sensing area of sensor in circle. From Fig. 21.1. it is observed that there is same area i.e. overlap areas cover by different sensors and these sensors forming the different fields [23]. Figure 21.2. The maximum disjoint cover set number (*T*) depend on the parameter like sensor's location, their sensing ranges, total number of sensors & size of target area. To cover the complete target area all sensors must be activated all time.If not the target may not be cover and deployment may be fails [8]. As shown in Fig. 21.2 the three sensors forming the six different fields. When all of the sensors are activated and form the fields the upper limit of forming field *T* is denoted by $\tilde{T}$and can be estimated as the minimum number of sensors that cover a field in the target area as

$$\tilde{\mathrm{T}} = \min_{\mathrm{j}=1,2,\ldots\mathrm{nf}} (|\mathrm{Fj}|) \tag{21.2}$$

where Fj denotes the set of sensors that cover the field j, nF is the number of fields formed by all of the sensors. In the case of Fig. 21.2, the values of $|F1|$, $|F2|$, … $|F6|$ are 1, 2, 1, 2, 3, 2, respectively, so the value of $\tilde{T}$ is 1.

The K-coverage problem is known as to find maximum number of disjoint complete sets i.e. Calculation of upper limit of T i.e. $\tilde{T}$ [6], which can be addressed by polynomial-time algorithms. Because there are fields only covered by $\tilde{T}$ sensors, the maximum number of disjoint complete cover sets is no larger than $\tilde{T}$ (i.e., $T \leq \tilde{T}$) [8].

For the optimal coverage the whole area cover by the minimum number of sensor nodes given by the equation

$$\frac{\text{Total Area}}{N \times r^2\pi} = \frac{2\pi}{\sqrt{27}} \tag{21.3}$$

The minimum number of sensor node is denoted by N and r is the sensing range of the node which is very small as compare to

As all of the sensors have the same sensing range R hence in a successful deployment, we have

$$\eta = \frac{N\pi R^2}{\breve{T}LW} \geq \frac{2\pi}{\sqrt{27}} \approx 1.21 \tag{21.4}$$

Where η is termed the redundancy rate.

(b) Calculation of the Coverage Percentage

It is very difficult to calculate coverage percentage directly to calculate the coverage percentage of the total sensing area simply divided the target area into grid and area inside the grid consider as covered area as shown in Fig. 21.3. For example (as Fig. 21.3), in a target area there are three active sensors s1, $s2$ & $s3$, which is composed of grids with a size d. The sensors' circular sensing areas are denoted as $A1$, $A2$, and $A3$. The numbers of grids covered by the three sensors are counted as 21, 21, and 20, respectively. The total number of grids covered by the sensors is 44. Therefore [8], the coverage percentage of the sensors to the target area with 13×8 grids is approximately $\frac{44}{3\times 8} \approx 42.3\ \%$ although the estimated coverage percentage is smaller than the actual coverage percentage, there are no blind points in the target area when the estimated coverage percentage is 100 %.

The three circular sensing areas $A1$, $A2$, $A3$ are covered by $s1$, $s2$, and $s3$ respectively. The number in grids indicates their field indexes. Note that the grids contained in the same field are covered by the same set of sensors the indexes of fields have also been exemplified in Fig. 21.3. The grids in the same field are covered by the same set of sensors. The higher the resolution of grids is, the higher the accuracy is for representing fields in the target area. In order to approximate fields better, the grid size is selected based on the sensors' sensing ranges. After determining the coverage percentage of each field, the fields can be regarded as targets [8] and can be used for computing the coverage percentage of sensors instead of using grids.

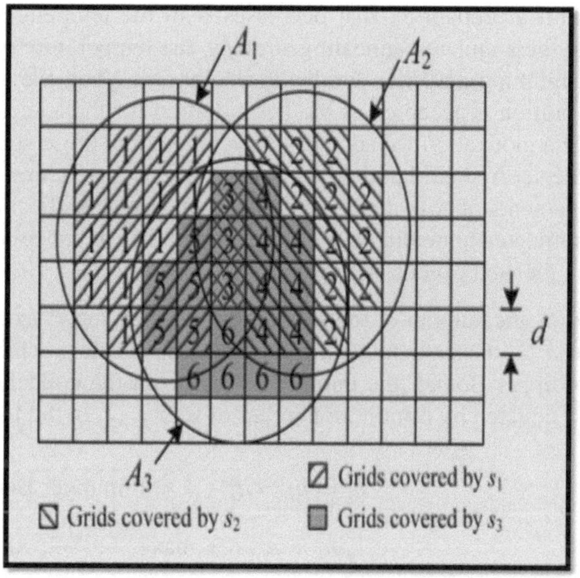

Fig. 21.3 Example for evaluating the coverage percentage of sensors to an area by dividing the target area into grids with a size *d*

21.4 Results Analysis and Discussion

In this section we describe the implementation, the simulation environment and the simulation results of the heuristic. With this algorithm we solved the SET k-cover problem by find out the maximum disjoint complete set. A. Simulated Annealing – Parameters. The performance of the algorithm is measure by approximate measurement of efficiency of algorithm. The sensors are randomly deployed in the field. At the beginning of the algorithm choose the clusterhead among these sensors. Any sensor become the clusterhead. The selection of the clusterhead is depend on the distance from the base station and energy level of the sensor. After choosing the clusterhead the sensors randomly joins the clusterhead and become the member of clusterhead. After forming the groups the algorithm forms the complete cover set such that members of each group cover complete target area and the cost function of this subcollection is:

$$\mathrm{CF} = a \cdot \sum_{\mathrm{i}=1}^{kj} 1 + b \cdot \sum_{i-k+1}^{nj} \frac{1}{mi} \tag{21.5}$$

Where the subcollections $i = 1,\ldots, kj$ are those that cover all elements from A. For the subcollections $i = \mathrm{k}_{\mathrm{j+1}}, \ldots, n$, that do not cover all elements from A, the number of uncovered elements from A denoted as nz, is calculated. If a new solution has a higher value of the cost function than the current solution, the new solution is accepted. If the value of the cost function for the new solution is lower than the value for the current solution, the new solution is accepted as a current solution with a probability that decreases with the temperature of the system. According to the simulated annealing strategy, the temperature of the system is highest at the beginning and then gradually decreases. When the system cools down, the final solution is accepted as the best solution for that execution of simulated annealing. The motivation for such cost function is that those subcollections that do not cover the set A should add to the value of the cost function depending on the number of elements of A that are not covered by that subcollection. After an execution of the simulated annealing algorithm, the number of subcollections n_{J+1} that is attempted in the next execution depends on:

- k_j, the number of the subcollections from the jth execution that cover the set A,
- k_{max}, the maximum number of the subcollections achieved in all previous executions,
- upper Bound, the number of sets from the collection C containing the element covered by minimum number of sets, e_{min}, in the following way:

$$kj = nj = nj + 1 = \text{minupper Bound2}.nj$$
$$k_j\ n_j = n_{j+1} = \max\left(\frac{kj + nj}{2},\ \text{kmax} + 1\right) \tag{21.6}$$

If the number of the subcollections covering A is equal to nj, the next execution attempts to achieve higher number of the subcollections that cover A, while in the case where kj is smaller than nj, the attempted number of subcollections for the next execution of simulated annealing is decreased, but never below an already achieved best result k max. The simulation stops if afterlimit number of attempts k, the achieved number of subcollections covering A, does not change.

(a) **Simulation Results**

Figure 21.4 shows the senior of wireless sensor network having 100 sensors with ten targets and one base station randomly deployed in a field. The beginning of the algorithm first cluster head is to be selected. All sensors not to be communicated with the base station directly but the data will send to the basestation through the clusterhead. So the clusterhead communicate to the basestation. Any sensor becomes a clusterhead the selection of the clusterhead is depending on the energy level of the clusterhead and the distance from the basestation.

Figure 21.5 shows the data is to be transmitted to the basestation. The sensor sense the data from the environment after sense the data, send to process the data and after processing the data the basestation.The graph shows the throughput versus time. The throughput increases with time the graph Fig. 21.7 shows the Avg throughput of the sensor network. On x axis shows the total coverage of the network and y axis shows the Avg throughput from the graph it is observed that as the throughput increases up to 30 meters but as the coverage increases above 30 meters the throughput decreases (Figs. 21.6, 21.8).

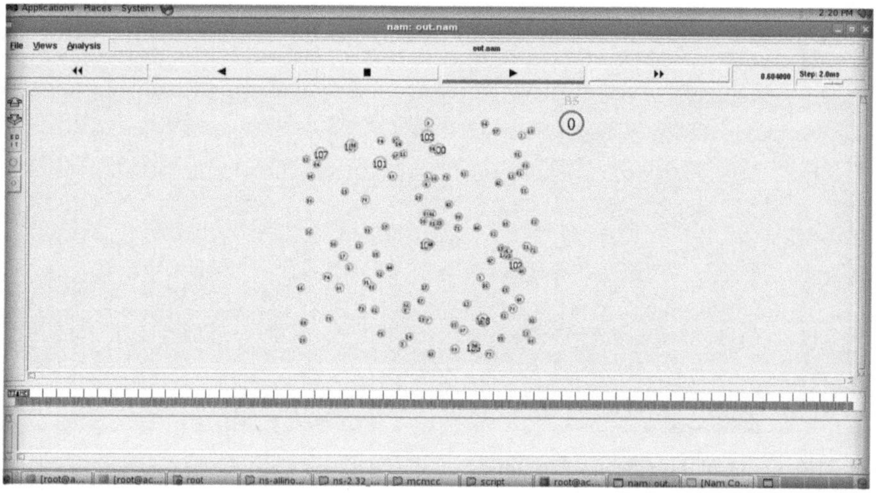

Fig. 21.4 A set of 100 sensors with 10 targets and one base station deploy randomly in a field

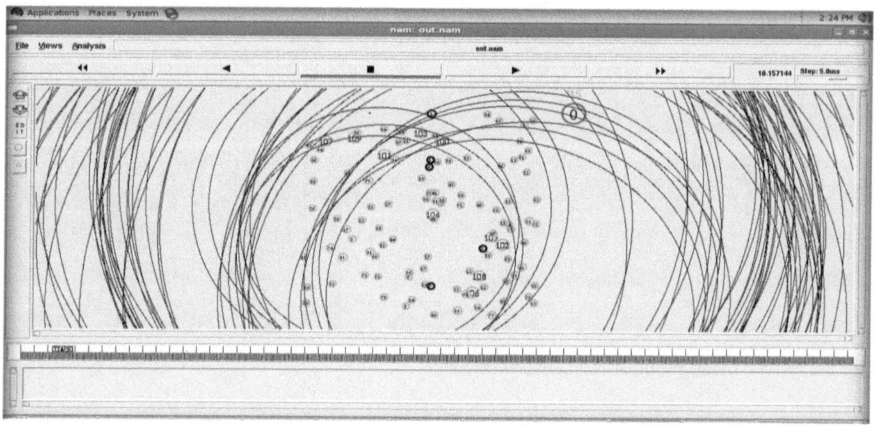

Fig. 21.5 Communication in sensors network

The above graph shows the average energy consume by the sensor network on x axis we take the sensing range and on y axis we take the Avg energy consume from the graph it shows that for 100 sensors the Avg energy consume of the network is between 15 and 40 joule for different sensing ranges (Fig. 21.9).

The graph shows the form cover set versus coverage from the graph it is observed that initially the cover set from maximum but as the sensing range of the sensor increases gradually decrease the formed cover set up to 30 meters and above 30 meters it increased again .

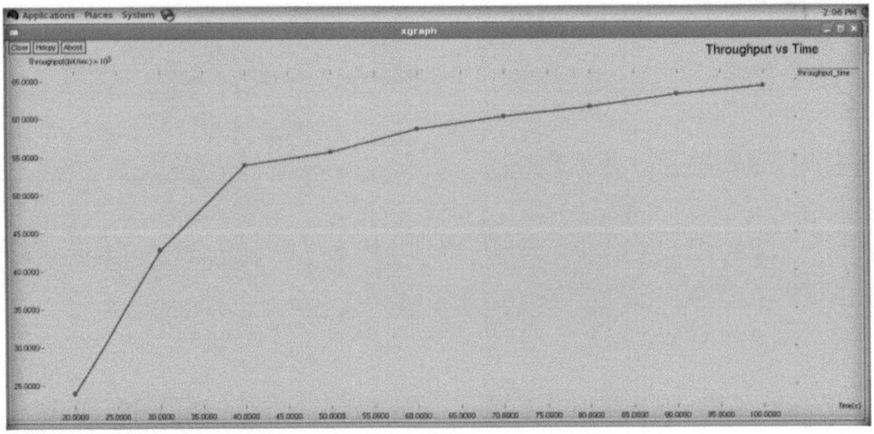

Fig. 21.6 Graph of throughput versus time

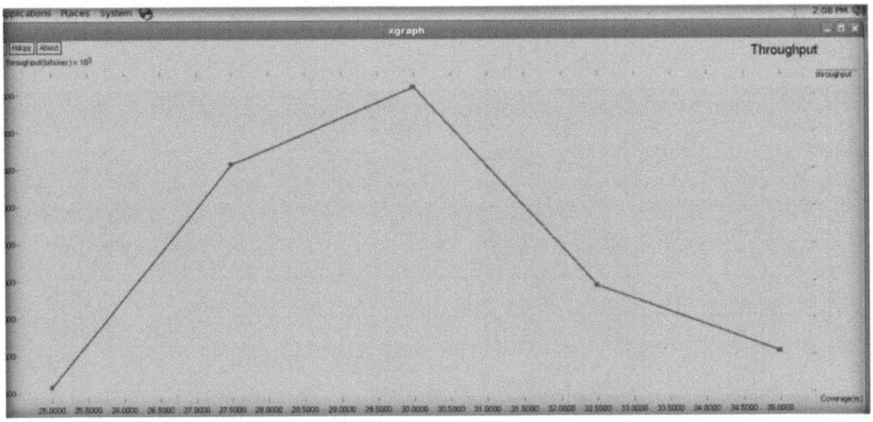

Fig. 21.7 Avg throughput of the sensor network

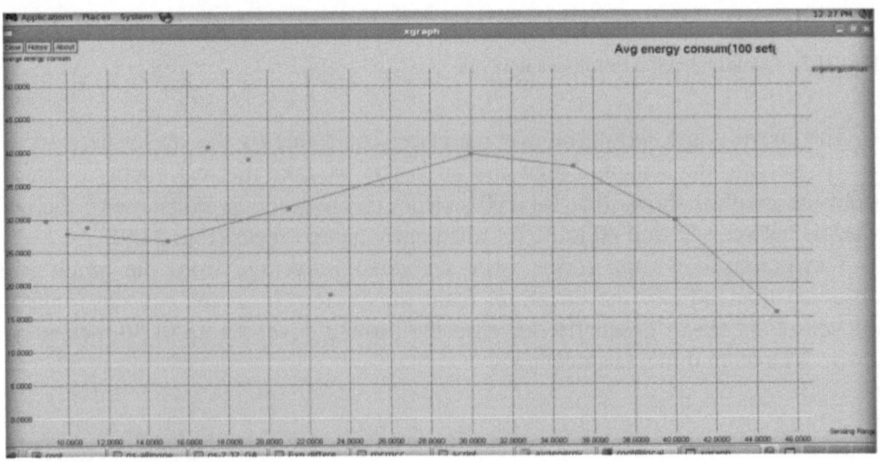

Fig. 21.8 Avg energy consume by sensor network

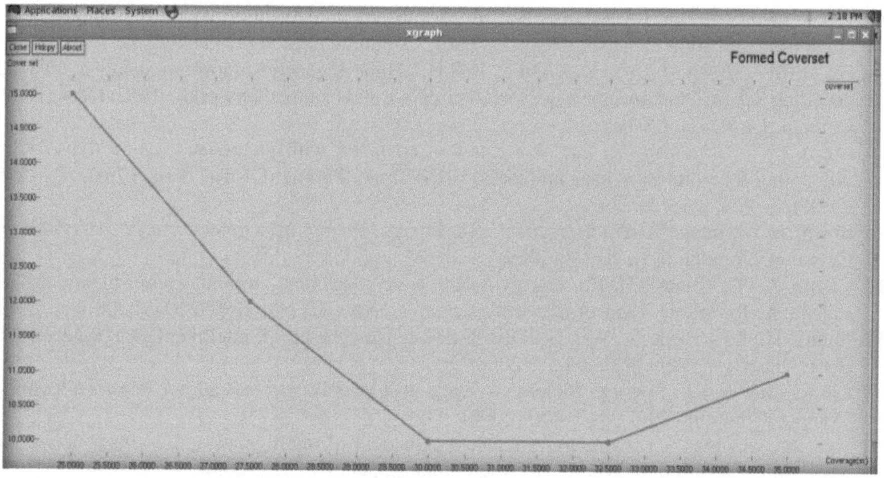

Fig. 21.9 Formed cover set versus coverage of sensors

21.5 Conclusion

We efficiently monitor the physical environment with wireless sensor network. As sensor nodes are battery operated means there is limited energy in a sensor node hence to build energy efficient network is main concern in wireless sensor network. In order to maximizing the lifetime of wireless sensor network nodes are divided into disjoint complete sets. Once we form disjoint complete set we activated only one set at a time and remaining are deactivated. We save energy of the network with the scheme. Our future work will explore the possible strategies of sensor networks deployment that will ensure total coverage of the monitor area.

References

1. Younis, M., Akkaya, K.: Strategies and techniques for node placement in wireless sensor networks: a survey. Ad Hoc Netw. **6**(4), 621–655 (2008)
2. Tian, D., Georganas, N.D.: Connectivity maintenance and coverage preservation in wireless sensor networks. Ad Hoc Netw. **3**(6), 744–761 (2005)
3. Zhangm, H., Hou, J.C.: Maintaining sensing coverage and connectivity in large sensor networks. Ad Hoc Sens. Wireless Netw. **1**(1–2), 89–124 (2005)
4. Cardei, M., Wu, J.: Coverage in wireless sensor networks. In: Ilyas, M., Mahgoub, I. (eds.) Handbook of Sensor Networks: Compact Wireless and Wired Sensing Systems, pp. 432–446. CRC Press, Boca Raton (2004)
5. Baek, S.J., de Veciana, G., Su, X.: Minimizing energy consumption in large-scale sensor networks through distributed data compression and hierarchical aggregation. IEEE J. Sel. Areas Commun. **22**(6), 1130–1140 (2004)
6. Wang, X., Xing, G., Zhang, Y., Lu, C., Pless, R., Gill, C.: Integrated coverage and connectivity configuration in wireless sensor networks. In: Proceedings of 1st International Conference on Embedded Networked Sensor Systems, pp. 28–39. Los Angeles, CA, 2003

7. Lin, F.Y.S., Chiu, P.L.: A near-optimal sensor placement algorithm to achieve complete coverage/discrimination in sensor networks. IEEE Commun. Lett. **9**(1), 43–45 (2005)
8. Hu, X.M., Zhang, J., Yu, Y., Chung, H.S.H.: Hybrid genetic algorithm using a forward encoding scheme for lifetime maximization of wireless sensor networks. IEEE Trans. Evol. Comput. **14**(5), 766 (2010)
9. Liu, C., Wu, K., Xiao, Y., Sun, B.: Random coverage with guaranteed connectivity: joint scheduling for wireless sensor networks. IEEE Trans. Parallel Distrib. Syst. **17**(6), 562–575 (2006)
10. Shwe, H.Y., Jiang, X.-H., Horiguchi, S.: Energy saving in wireless sensor networks. J. Commun. Comput. **6**(5), 20–28 (2009)
11. Chang, C.-Y., Chang, H.-R.: Energy-aware node placement, topology control and MAC scheduling for wireless sensor networks. Comput. Netw. **52**(11), 2189–2204 (2008)
12. Leung, H., Chandana, S., Wei, S.: Distributed sensing based on intelligent sensor networks. IEEE Circ. Syst. Mag. **8**(2), 38–52 (2008)
13. Cardei, M., Wu, J.: Energy-efficient coverage problems in wireless ad-hoc sensor networks. Comput. Commun. **29**(4), 413–420 (2006)
14. Heo, N., Varshney, P.K.: Energy-efficient deployment of intelligent mobile sensor networks. IEEE Trans. Syst. Man. Cybern. A, Syst. Humans **35**(1), 78–92 (2005)
15. Wang, Y.-C., Hu, C–.C., Tseng, Y.-C.: Efficient placement and dispatch of sensors in a wireless sensor network. IEEE Trans. Mob. Comput. **7**(2), 262–274 (2008)
16. Lin, J.-W., Chen, Y.-T.: Improving the coverage of randomized scheduling in wireless sensor networks. IEEE Trans. Wireless Commun. **7**(12), 4807–4812 (2008)
17. Iyengar, S.S., Wu, H.-C., Balakrishnan, N., Chang, S.Y.: Biologically inspired cooperative routing for wireless mobile sensor networks. IEEE Syst. J. **1**(1), 29–37 (2007)
18. Cui, S., Madan, R., Goldsmith, A.J., Lall, S.: Cross-layer energy and delay optimization in small-scale sensor networks. IEEE Trans. Wireless Commun. **6**(10), 3688–3699 (2007)
19. Yu, Y., Prasanna, V.K., Krishnamachari, B.: Energy minimization for real-time data gathering in wireless sensor networks. IEEE Trans. Wireless Commun. **5**(11), 3087–3096 (2006)
20. Wang, L., Xiao, Y.: A survey of energy-efficient scheduling mechanisms in sensor networks. Mob. Netw. Appl. **11**(5), 723–740 (2006)
21. Howard, A., Matari′c, M.J., Sukhatme, G.S.: An incremental self deployment algorithm for mobile sensor networks. Auton. Robots **13**(2), 113–126 (2002)
22. Abrams, Z., Goel, A., Plotkin, S.: Set k-cover algorithms for energy efficient monitoring in wireless sensor networks. In: Proceedings of 3rd International Symposium on Information Processing in Sensor Networks, pp. 424–432 (2004)
23. M. Cardei, D. MacCallum, M. X. Cheng, M. Min, X. Jia, D. Li, and D.-Z.Du, "Wireless sensor networks with energy efficient organization,"J. Interconnect.Newt, vol. 3, nos. 3–4, pp. 213–229, 2002
24. Slijepcevic, S., Potkonjak, M.: Power efficient organization of wireless sensor networks. In: IEEE International Conference on Communication, vol. 2, pp. 472–476, Finland, 2001
25. Berman, P., Calinescu, G., Shah, C., Zelikovsky, A.: Efficient energy management in sensor networks. In: Pan, Y., Xiao, Y. (eds.) Ad Hoc and Sensor Networks, Wireless Networking and Mobile Computing, pp. 71–90. Nova Science Publishers, New York (2005)
26. Zhang, H., Hou, J.C.: Maximising α-lifetime for wireless sensor networks. Int. J. Sens. Netw. **1**(1–2), 64–71 (2000)

Chapter 22
A Secure and Efficient Password-Based User Authentication Scheme for Hierarchical Wireless Sensor Networks

Chandra Sekhar Vorugunti and Mrudula Sarvabhatla

Abstract Wireless sensor networks (WSN) have attracted increasing usage in critical data centric applications like battle field surveillance, medical, ocean and, other real-time applications. External users are generally interested in accessing real-time aggregated data, directly from the desired sensor nodes, inside WSN, when needed. User authentication is a fundamental issue in these scenarios. It is of great challenge to check user authenticity, protects the users and system security, and privacy from malicious adversaries. In 2012, Das et al. proposed a dynamic password-based authentication scheme for WSN and claimed that their scheme provides better security compared to other related schemes and resists masquerade, replay, guessing, impersonation attacks. In this paper, we will show that the Das et al. scheme cannot confront any of the attacks they claimed that their scheme will resist. As a part of our contribution, we propose an efficient and secure password-based authentication scheme, for WSN, which resists all the major security attacks and requires very light storage and computational cost compared to Das et al. and other relevant schemes, while preserving their merits.

Keywords WSN security • Authentication • Wireless network security • Smart cards

C. S. Vorugunti (✉)
Dhirubhai Ambani Institute of Information and Communication Technology,
Gandhi Nagar 382007, Gujarat, India
e-mail: vorugunti_chandra_sekhar@daiict.ac.in

M. Sarvabhatla
Sri Venkateswara University, Tirupathi 517502, Andhra Pradesh, India
e-mail: mrudula.s911@gmail.com

R. Maringanti et al. (eds.), *Proceedings of Ninth International Conference on Wireless Communication and Sensor Networks*, Lecture Notes in Electrical Engineering 299,
DOI: 10.1007/978-81-322-1823-4_22, © Springer India 2014

22.1 Introduction

Wireless sensor network (WSN) is a distributed heterogeneous system combining tiny sensors (varies from a count of hundreds to thousands), actuators, and low power nodes with limited computation, communication, and storage capabilities to sense and process the real-time data.

Generally WSNs are used to collect and process critical data like battle field monitoring, seismic activities, forest fire, etc. The data collected can be used to detect certain events like earth quake, bush fire, etc. and to trigger certain remedy activities. Usually, WSNs are deployed in remote places and are left unattended. Due to this the sensor nodes are prone to various security attacks like node capture, impersonation, denial of service, etc. by the adversaries which makes the critical data stored in the sensor obsolete.

The data collected and stored in a sensor node memory should be kept confidential and must be available only to the authorized people. Therefore, a WSN must be able to distinguish the legitimate users from illegitimate users.

Various researchers proposed protocols for secure authentication of users trying to access the sensor data [2–12]. Unfortunately, most of the protocols are analyzed insecure shortly, after they were put forward [3–12]. In 2012, Das et al. [1] proposed a dynamic password-based user authentication scheme for hierarchical WSN and claimed that their protocol achieves better authentication security, efficiency, and resists cryptographic attacks like replay attack, password guessing attack, password change attack, impersonation attack, smart card breach attack, etc.

The rest of the paper is organized as follows: In Sect. 22.2, a brief review of Das et al. scheme is given. Section 22.3, describes the security weakness of Das et al. scheme. In Sect. 22.4, our improved scheme is proposed and its security analyses are discussed in Sect. 22.5. The cost and security comparison of various similar password-based user authentication schemes for WSN are given in Sects. 22.6 and 22.7 provides the conclusion of the paper.

22.2 Review of Ashok Kumar Das et al. Protocol

In this section, we examine the dynamic password-based user authentication scheme proposed by Das et al. [1] for WSN in 2012. The scheme is composed of following phases: predeployment, postdeployment, registration, login, authentication, password change, and dynamic node addition phase. The notations used in Das et al. [1] and our proposed scheme are listed below:

U_i	User
BS	Base station
S_j	Sensor node
CHu	Cluster head "u"
PW_i	Password of user U_i

ID_i	Identity of user U_i
CIDu	Identifier of cluster head CH_u
$h(.)$	A secure one-way hash function
X_s	Secret information maintained by the base station
X_A	Secret information shared between user and base station
MKCHu	Master key for cluster head CH_u
y	A secret random number chosen by user
T_i	Time at which bas station received the registration request from user U_i
T_u	Time at which the cluster head CH_u is deployed into the WSN by base station
$A \| B$	Data A concatenates with data B
$A \oplus B$	XOR operation of A and B

22.2.1 Predeployment Phase

The base station (BS) performs the following steps in offline mode before the actual deployment of the sensor nodes and cluster heads in deployment field.

Step 1. The BS assigns a unique identifier CIDu to each cluster head CH_u which will be deployed in the target field. For each deployed regular sensor node S_i, the BS assigns a unique identifier, say SID_i.

Step 2. The BS then randomly selects a unique master key, say $MKCH_u$ for each cluster head CHu. The master key is known only to the cluster head CHu and the BS only. Similarly, the BS also assigns a unique randomly generated master key, MKS_i for each deployed regular sensor node S_i, which will be shared with the BS only.

Step 3. Finally, the BS loads the following information into the memory of each cluster head CHu ($u = 1, 2, \ldots, m$): (1) its own identifier, CIDu and (2) its own master key MKCHu. Each deployed regular sensor node S_i headed by the cluster head CHu is loaded with the following information: (1) its own identifier, SID_i and (2) its own master key MKS_i.

22.2.2 Registration Phase

This phase is invoked whenever a user U_i registers with the BS for the first time.

Step 1. The user U_i selects the identifier ID_i, a random number y, and the password PW_i. U_i then computes $RPW_i = h(y \| PW_i)$. U_i provides the computed masked password RPW_i and ID_i to the BS via a secure channel for registration.

Step 2. The BS computes the following variables for U_i. $f_i = h(ID_i \| X_s)$, $x = h(RPW_i \| X_A)$, $r_i = h(y \| x)$, and $e_i = f_i \oplus x = h(ID_i \| X_s) \oplus h(RPW_i \| X_A)$. The secret information X_S is only known to the BS. The secret information X_A is known to U_i and the BS.

Step 3. The BS then selects all "m" deployed cluster heads in the network CH_1, CH_2,...,CH_m, which will be deployed during the initial deployment phase, and computes the m key-plus-id combinations {(Ku, CIDu)| $1 \le u \le m$}, where Ku = E_{MKCHi} (ID_i||CIDu||X_s).

Step 4. For dynamic cluster head addition phase, the m' cluster heads, CH_{m+1}, CH_{m+2},..., $CH_{m+m'}$, will be deployed later after the initial deployment in the network in order to replace some compromised cluster heads, if any, and add some fresh cluster heads along with sensor nodes. For this purpose, the BS computes another m' key-plus-id combinations {(Ku + i, CIDu + i) |$1 \le i \le m'$}, where Ku + i = $E_{\text{MKCHi}+i}$ (ID_i||CIDu + i||X_s). CIDu + i is the unique identifier generated by the BS for the cluster head CHu + i to be deployed during the dynamic node addition phase and MKCHu + i the unique master key randomly generated by the BS for CHu + i, which is shared between it and the BS.

Step 5. Finally, the BS issues a tamper-proof smart card with the following parameters stored in it : (1) ID_i (2) y, (3) X_A, (4) r_i, (5) e_i, (6) $h(.)$, and (7) $m + m'$ key-plus-id combinations {(Ku, CIDu) | $1 \le u \le m + m'$}. The value of $m + m'$ is chosen according to memory availability of the smart card.

22.2.3 Login Phase

Whenever the user U_i wants to access real-time data from a sensor of a deployed WSN, the user U_i needs to perform the following steps:

Step 1. U_i inserts his/her smart card into the card reader of a specific terminal and provides his/her password PW_i.

Step 2. The smart card then computes the masked password of the user U_i as $RPW'_i = h(y||PW_i)$. Using the computed masked password, the smart card further computes $x' = h(RPW'_i||X_A)$ and $r_i' = h(y||x')$, and then checks whether $r_i' = r_i$, i.e., the computed r_i equals to the received r_i. If this verification fails, it means U_i entered the wrong password and the scheme terminates else the smart card proceeds to perform the following steps.

Step 3. Using the system's current timestamp T_1, the smart card computes $N_i = h(x'||T_1)$.

Step 4. The user U_i selects a cluster head, say CHu from which the real-time data can be accessed inside WSN. Corresponding to CHu, the smart card selects the encrypted master key of CHu, i.e., Ku from its memory and computes a cipher text message E_{Ki} (ID_i||CIDu||N_i||e_i||T_1). Finally, the user sends the message <ID_i||CIDu||E_{Ki} (ID_i||CIDu||N_i||e_i||T_1)> to the BS via a public channel.

22.2.4 Authentication Phase

On receiving the login request message $<\text{ID}_i||\text{CIDu}||E_{\text{Ki}}(\text{ID}_i||\text{CIDu}||N_i||e_i||T_1)>$ from the user U_i, the BS performs the following steps in order to authenticate the user U_i.

Step 1. The BS computes a key K using the stored master key MKCHu of the cluster head CHu as $\text{Ku} = E_{\text{MKCHi}}\ (\text{ID}_i||\text{CID}_i||\text{X}_\text{s})$. Using this computed key K, the BS decrypts $E_{\text{Ki}}(\text{ID}_i||\text{CID}_i||N_i||e_i||\text{T}_1)$, i.e.,

$$D_{\text{Ku}}[E_{\text{Ku}}(\text{ID}_i||\text{CID}_u||N_i||e_i||T_1)] = (\text{ID}_i||\text{CIDu}||N_i||e_i||T_1).$$

Step 2. The BS checks, if retrieved ID_i is equal to received ID_i and also if retrieved CIDu is equal to received CIDu. If these holds, the BS further checks if $|T_1 - T_1^*| < \Delta T_1$, where T_1^* is the current system time stamp of the BS and ΔT_1 is the expected time interval for the transmission delay. Now if it holds, the BS further computes $X = h(\text{ID}_i||X_s)$, $Y = e_i \oplus X$, and $Z = h(Y||T_1)$. If $Z = N_i$, then the BS accepts U_i's login request and U_i is considered as a valid user by the BS. Otherwise, the scheme terminates.

Step 3. Using the current system time stamp T_2, the BS computes $u = h(Y||T_2)$ and produces a cipher text message encrypted using the master key MKCHu of the cluster head CHu as $E_{\text{MKCHi}}(\text{ID}_i||\text{CID}_i||u||T_1||T_2||X||e_i)$. The BS sends the message $<\text{ID}_i||\text{CIDu}||E_{\text{MKCHi}}(\text{ID}_i||\text{CIDu}||u||T_1||T_2||X||e_i)>$ to the corresponding cluster head CH_j.

Step 4. After receiving the message in Step 3 from the BS, the cluster head CHu decrypts $E_{\text{MKCHi}}(\text{ID}_i||\text{CIDu}||u||T_1||T_2||X||e_i)$ using its own master key MKCHu as: $D_{\text{MKCHi}}[E_{\text{MKCHi}}(\text{ID}_i||\text{CIDu}||u||T_1||T_2||X||e_i)] = (\text{ID}_i||\text{CIDu}||\text{u}||T_1||T_2||X||e_i)$. CHu then checks if retrieved ID_i is equal to received ID_i and also if retrieved CIDu is equal to received CIDu. If these holds, CHu further checks if $|T_2 - T_2^*| < \Delta T_2$, where T_2^* is the current system time stamp of the CHu and ΔT_2 is the expected time interval for the transmission delay. If it holds good, CHu computes $v = e_i \oplus X = h(\text{RPW}_i||X_A)$, $w = h(v||T_2) = h(h(\text{RPW}_i||X_A)||T_2)$. CHu then checks if $w = u$. If it does not hold, the scheme terminates. Otherwise if it holds, the user U_i is considered as a valid user and authenticated by CHu. CHu computes a secret session key SK shared with the user U_i as $\text{SK} = h(\text{ID}_i||\text{CIDu}||e_i||T_1)$. Finally, CHu sends an acknowledgment to the user U_i via other cluster heads and the BS and responds to the query of the user U_i.

Step 5. After receiving the acknowledgment from CHu, the user U_i computes the same secret session key shared with CHu using its previous system time stamp T_1, ID_i, CIDu, and e_i as $\text{SK} = h(\text{ID}_i||\text{CIDu}||e_i||T_1)$. Thus, both user U_i and cluster head CHu will communicate securely in future using the derived secret session key SK.

22.2.5 *Dynamic Node Addition Phase*

In dynamic node addition phase, if any of the existing sensor nodes or cluster heads are captured by an adversary or expired due to energy depletion, then new nodes need to be added into the network in order to replace those nodes. Unique identifier and randomly generated unique master key are assigned to a sensor node, cluster head, and loaded into their memory.

22.3 Analysis of Weakness of Das et al. Scheme

22.3.1 *Huge Data Storage and Computation Requirement for Generating User Smart Card*

In Das et al. scheme, the smart card memory is stored with key-plus-id combination (Ku, CIDu) of all cluster heads in the WSN. Based on the Das et al. discussion, for a total of 22,000 sensor nodes to be deployed and if each cluster head can handle 220 sensor nodes then there are total $m = 100$ cluster heads needed and for dynamic cluster head addition $m' = 100$ cluster heads are reserved. So a total of $m + m' = 100 + 100$\{Ku, CIDu\} details are stored, where Ku $= E_{\mathrm{MKCHi}}$ \{ID_i,CIDu, X_s\}. A total of 200 encryptions need to be performed for each user smart card. If the system contains n users, then a total of (n * 200) encryptions need to be performed to load the smart card memory of corresponding user which requires huge computation cost from the BS. The major issue is that the user may not interested or in need of data from all the cluster heads. Hence storing all the m + m' cluster head details is a major drawback in Das et al. scheme.

22.3.2 *Smart Card Breach Attack*

If a legal adversary E stolen the smart card of a valid user U_i and performed the smart card breach attack, as mentioned in [18]. E will come to know the identity ID_i of the user, y, X_A, r_i, e_i, $h(.)$ and all the key- plus-Id combinations {(Ku, CIDu)} of the cluster heads. (The adversary E can note down all the key-plus-id combinations for further attacks). There is a data leakage of all the keys of cluster heads which are used for encrypting the login request message to BS. Hence, Das et al. scheme suffers from a severe security threat of data leakage of critical encrypting keys to the outside world. On performing the smart card breach attack, the adversary E is revealed with the identity ID_i of the user. Hence, Das et al. scheme also suffers from a drawback of failure to preserve user anonymity which is critical requirement in insecure WSN.

22.3.3 Password Guessing Attack

If a legal adversary E attains the smart card of a legal user U_i and retrieved the data in it, the adversary "E" can have the following values: y, X_A, $r_i = h(y||h(\mathrm{RPW}_i||X_A)) = h(y||h(y||\mathrm{PW}_i||X_A))$. The adversary E knows all the values in r_i except PW_i. By using the uniformly distributed dictionary, E can perform password guessing attack on r_i. Guess a secret value PW* and check if $r_i = h(y||h(y||\mathrm{PW}_i^*||X_A))$. If they are equal then the U_i password is PW*. Otherwise, E can repeat the process to get correct value PW*. Hence, in Das et al. scheme the legal user password can be revealed to the adversary. By knowing the ID_i and PW_i of a legal user, the adversary can perform all major cryptographic attacks.

22.3.4 User Impersonation Attack

In Das et al. scheme the smart card sends the login request message <ID_i, CIDu, $\mathrm{E_{Ki}}(\mathrm{ID}_i$, CIDu, N_i, e_i, $T_1)$> to the BS. As discussed in 3.2, the adversary E can have ID_i, y, X_A, r_i, e_i, $h(.)$ and all the key-plus-id combinations {(Ku, CIDu)} of the cluster heads. Once ID of the user matches, the adversary E can intercept the login request sent by U_i to BS and can decrypt the login message by retrieving Ku from the Key-plus-id pairs. On decrypting the login message, i.e., $D_{\mathrm{Ki}}(E_{\mathrm{Ki}}(\mathrm{ID}_i, \mathrm{CID}_i, N_i, e_i, T_1))$, E gets ID_i, CIDu, N_i, e_i, T_1. Now, E can frame the following values $\mathrm{RPW} = h(y||\mathrm{PW}_i)$, $x = h(\mathrm{RPW}||X_A)$. $N_i^* = h(x||T_1^*)$. Now E can frame a new valid login request <ID_i, CIDu, $\mathrm{E_{Ki}}(\mathrm{ID}_i$, CIDu, N_i^*, e_i, $T_1^*)$> using the new time stamp T_1^*. The new login message will sure pass the authentication process of BS in Step 2. Therefore, the adversary E can successfully impersonate U_i by replaying the login request message. Hence Das et al. scheme suffers from user impersonation attack and replay attack.

22.3.5 Framing the Session Key Between the Legal User and Cluster Head by the Adversary

In Das et al. scheme, once the user is authenticated by the cluster head the session key will be framed between the cluster head and the user as: $\mathrm{S.K} = h(\mathrm{ID}_i||\mathrm{CIDu}||e_i||T_1)$. An adversary E once written down all the smart card entries and key-plus-id combinations of U_i in a paper and intercepted the login message sent by smart card as discussed in Sect. 22.3.3, this section will have ID_i, CIDu, N_i, e_i, T_1. With these values, E can easily frame the session key $\mathrm{S.K} = h(\mathrm{ID}_i||\mathrm{CIDu}||e_i||T_1)$, as E knows all the values required to frame SK. Hence an adversary can read, alter all the data exchanged between user and the cluster head on knowing the session key. Hence Das et al. scheme suffers from the biggest drawback of framing the session key by the adversary and can intercept all the data exchanged between the user and the cluster head.

22.4 Our Proposed Authentication Protocol

In this section, we present our improved dynamic password-based user authentication scheme for hierarchical WSN as a remedy to Das et al. scheme while preserving their merits.

22.4.1 *Registration Phase*

This phase is invoked whenever a user U_i wants to register first time. The following steps are performed:

Step 1. User U_i selects ID_i, PW_i and a random number y, then calculates $\mathrm{RPW}_i = h(y||\mathrm{PW}_i)$ and sends ID_i, RPW_i, y to the smart card SC in a secure channel.

Step 2. On receiving the registration request from U_i at T_i, the BS computes the following: $f_i = h(\mathrm{ID}_i||X_s||T_i)$, $x = h(y||\mathrm{RPW}_i||T_i)$, $r_i = h(T_i||x||y)$, $u = h(T_i||y||\mathrm{ID}_i)$, $\mathrm{TID}_i = (\mathrm{ID}_i||h(T_i))$, $e_i = u \oplus x$.

Step 3. The BS issues the smart card containing $\{\mathrm{TID}_i, (y||T_i), r_i, e_i, h(.)\}$ to U_i in a secure manner.

22.4.2 *Login Phase*

Step 1. Whenever U_i wants to login into the system to access the sensor node data, U_i inserts his smart card into the card reader and enters y, PW_i.

Step 2. The smart card will perform following actions: From the entered value "y," retrieve T_i from $(y||T_i)$ and computes the masked password $\mathrm{RPW}' = h(y||\mathrm{RPW}_i)$.

Step 3. Compute: $x = h(y||\mathrm{RPW}_i||T_i)$.

Step 4. Compute $r_i = h(T_i||x||y)$ and verifies whether r_i computed equals r_i received. If not then the smart card rejects the login request else U_i is authenticated and smart card proceeds by computing $N_i = h(\mathrm{ID}_i||x||T_i||T_1)$, $\mathrm{TID}_{i1} = (\mathrm{TID}_i||h(T_1))$.

Step 5. The user U_i selects the cluster head CIDu from which the real-time data can be accessed inside WSN. Smart card further computes $X_a = h(\mathrm{ID}_i||T_1||y||T_i||\mathrm{CIDu})$, $\mathrm{TCIDu} = (\mathrm{CIDu}||T_1)$.

Step 6. S.C frames the encrypted login request message $E_{Xa}\{\mathrm{ID}_i, \mathrm{CIDu}, N_i, e_i, T_i, T_1\}$ and finally sends the message <TID_{i1}, TCIDu, $E_{Xa}\{\mathrm{ID}_i, \mathrm{CIDu}, N_i, e_i, T_i, T_1\}$> to BS.

22.4.3 Authentication Phase

On receiving the login request from the smart card, the B.S performs the following steps.

Step 1. Receive and intercept the message: <TID_{i1}, TCIDu, $E_{\text{Xa}}\{\text{ID}_i$, CIDu, N_i, e_i, T_i, $T_1\}$>.
Step 2. B.S searches among the list of cluster heads for CIDu which will have a match with TCIDu as TCIDu = (CIDu||T_1).
Step 3. On identifying the CIDu, the BS retrieves T_1 from (CIDu||T_1) and computes $h(T_1)$.
Step 4. B.S gets TID_i from (TID_i||$h(T_1)$) and retrieves ID_i and $h(T_i)$ from TID_i.
Step 5. B.S computes $X_a = h(\text{ID}_i||T_1||y||T_i||\text{CIDu})$ and decrypts the login message, i.e., $D_{\text{Xa}}\ \{E_{\text{Xa}}\{\text{ID}_i$, CIDu, N_i, e_i, T_i, $T_1\}\}$ to get ID_i, CIDu, N_i, e_i, T_i, T_1.
Step 6. B.S further computes $x = e_i \oplus h(T_i||y||\text{ID}_i)$, $N_i = h(\text{ID}_i||x||T_i||T_1)$. If N_i computed = N_i received, user is authenticated by BS.
Step 7. B.S further computes $P = h(N_i||T_2||T_i)$ and frames an encrypted message $E_{\text{MKCHi}}\{\text{ID}_i$, CIDu, e_i, u, P, T_i, T_1, $T_2\}$ and sends it to cluster head CH_i.

Authentication by Cluster Head

Step 8. Receive and decrypt the login message $D_{\text{MKCHi}}\{E_{\text{MKCHi}}\{\text{ID}_i$, CIDu, e_i, u, P, T_i, T_1, $T_2\}\}$ to get ID_i, CIDu, e_i, u, P, T_i, T_1, T_2.
Step 9. C.H frames $e_i \oplus u = x$ and computes $N_i = h(\text{ID}_i||x||T_i||T_1)$.
Step 10. C.H further computes $P = h(N_i||T_2||T_i)$ and verifies computed $P = P$ received. If yes, user is authenticated and cluster head proceeds further else the request is rejected and the connection is closed.
Step 11. C.H frames the Session Key S.K = $h(\text{ID}_i||\text{CIDu}||T_i||T_2||e||T_1||T_i)$.
Step 12. Computes a cipher text message $E_i\{T_2$, T_1, T_c, ID_i, CIDu$\}$ and sends it to U_i.
Step 13. On receiving the acknowledgement, U_i decryts the message as U_i knows u, i.e., $D_i\{E_i\{T_2, T_1, T_i, \text{ID}_i, \text{CIDu}\}\}$ to get T_2, T_1, T_i, ID_i, CIDu.
Step 14. U_i frames the se,ssion key SK, $h(\text{ID}_i||\text{CIDu}||T_i||T_2||e||T_1||T_i)$. All the further communication between the user and cluster head is done by encrypting the message with the SK.

22.4.4 Dynamic Node Addition Phase

In our proposed scheme, if some sensor nodes or cluster heads are compromised or sensor node expired due to energy depletion then new nodes will be added to replace those nodes. Similar to Das et al. scheme in our scheme also the BS assigns new ID and master key to sensor node, cluster head, and loads the information into their memory. The BS updates the user U_i about the IDs of the new cluster heads added. While logging in, the user has to submit only the identity of cluster head from which the real-time data is needed. The SC and BS will take care of dynamic key generation and authentication. Thus, we completely eliminated the dependency of WSN structure on smart card key-plus-id entries, which leads to critical data leakage and decryption of login messages by the adversary.

22.5 Security Analysis

22.5.1 *Light Weight Data Storage and Computations for Generating User Smart Card*

In Das et al. scheme the smart card memory is stored with key-plus-id combination (Ku, CIDu) of all cluster heads in the WSN. If a smart card breach attack is done by the adversary, then the critical encrypting keys will be revealed to the adversary. In our protocol, we are not storing any key-plus-id combinations of cluster heads in the smart card. Once the user is logged in and authenticated by the smart card, then the smart card frames $X_a = h(\mathrm{ID}_i \| T_1 \| y \| T_i \| \mathrm{CIDu})$, which acts as an encrypting key where T_i is the time at which the user sent the registration request to the BS and T_1 is the time at which the smart card received the login request. Due to T_1, the encrypting key X_a became purely random and will be useful only for that session. The smart card computes the encrypting key only when there is requirement from the user to access the real-time data from a particular cluster head. So a huge computation cost and storage cost is reduced compared to Das et al. scheme.

22.5.2 *Resisting Smart Card Breach Attack*

If a legal adversary E stolen, the smart card of a valid user U_i and performed the smart card breach attack, as mentioned in [18], E can extract the values TID_i, $(y\|T_i)$, r_i, e_i, $h(.)$ where $\mathrm{TID}_i = (\mathrm{ID}_i \| h(T_i))$, $r_i = h(T_i \| x \| y)$, $e = u \oplus x$ where $u = h(T_i \| y \| \mathrm{ID}_i)$, $x = h(y \| \mathrm{RPW}_i \| T_i)$. As shown in the Table 22.1, an adversary does not know the ID_i of the user U_i because we are not storing plain ID_i in the smart card as done in Das et al. scheme [ID_i is concatenated with $h(T_i)$ to form TID_i]. As the adversary does not know the random number chosen by U_i, i.e., "y" (we are not storing "y" in smart card as done in Das et al. scheme), it is not possible for E to intercept T_i from $(y\|T_i)$ to frame $h(T_i)$ and to get ID_i from TID_i. Without knowing y, T_i, ID_i, it is not possible for E to frame x, r_i, u. Hence in our scheme, no critical data of legal user U_i is revealed to an adversary and also U_i's real identity, i.e., ID_i. Hence, our scheme preserves user anonymity and resists critical data leakage.

Table 22.1 Types of users and the values known to them

Type of user	Values known to the user	Values not known to the user
Legal user (U_i)	A legal user knows his own y, PW_i, ID_i, T_i	T_2, T_1, T_i
Legal adversary (E)	TID_{i1}, TCIDu of U_i	y, PW_i, ID_i, T_i of U_i

22.5.3 Resisting Password Guessing Attack

If a legal adversary E gets the smart card of user U_i and retrieved the data in it, the adversary "E" can have the following values: TID_i, $(y\|T_i)$, r_i, e_i, $h(.)$ where $\text{TID}_i = (\text{ID}_i\|h(T_i))$, $r_i = h(T_i\|x\|y)$, $e = u \oplus x$ where $u = h(T_i\|y\|\text{ID}_i)$, $x = h(y\|\text{RPW}_i\|T_i)$. The adversary E must frame "u" to get "x" from "e." As discussed above E does not know "y," "T_i," "ID_i," therefore it is not possible for E to frame "u" and compute "x" which contains RPW. Hence, in our scheme it is not possible for an adversary E to guess password of a legal user U_i and subsequently perform password change attack.

22.5.4 Resisting User Impersonation Attack

In our scheme, the smart card sends the login request message <TID_{i1}, CIDu, $E_{\text{Xa}}\{\text{ID}_i$, CIDu, N_i, e_i, T_i, $T_1\}$> to the BS where $\text{TID}_{i1} = (\text{TID}_i\|h(T_1))$, $\text{TCIDu} = (\text{CIDu}\|T_1)$. We are storing TID_i in the smart card of the user. While sending the login request, the SC sends TID_{i1}, CIDu. To get TID_i, the smart card must know T_1. To get T_1 the smart card must decrypt the login message using X_a. To frame X_a, the adversary must know ID_i, x, T_i, T_1. Hence, it is not possible for an adversary to frame the key X_a. Therefore, the adversary cannot decrypt the message and modify the message content. Hence, in our scheme it is not possible for any kind of adversary to impersonate the valid user U_i by replaying the login request messages. As discussed, we are not storing the real identity ID_i in the U_i smart card. While sending login request, the SC sends TID_{i1}. We have shown that it is impossible to find TID_i from TID_{i1} and ID_i from TID_i. Hence our scheme preserves user anonymity.

22.5.5 Impossible to Frame the Session Key Between the Legal User and Cluster Head by the Adversary

In our scheme, once the user is authenticated by the cluster head, a session key will be framed between the cluster head and the user where $\text{S.K} = h(\text{ID}_i\|\text{CIDu}\|T_i\|T_2\|e\|T_1\|T_i)$. As discussed, the adversary E does not know any of the values ID_i, CIDu, T_i, T_2, T_1, T_i except "e." Therefore, it is not possible for any kind of user to frame the session key. Hence, in our scheme data exchange between the cluster head and user is completely secure.

Table 22.2 Comparison of computational cost in different phases between the proposed scheme and the other schemes

Phase	User or node	Wong et al. (2006)	Das (2009)	Nyang and Lee (2009)	Huang et al. (2010)	He et al. (2009)	Vaidya et al. (2010)	Fan et al. (201 0)	Chen and Shih (2010)	Das et al. (2012)	Proposed
Registration (assuming 22,000 sensors to be deployed where a cluster head can handle 220 clusters)	User (U_i)	–	–	–	–	t_h	t_h		–	$1t_h$	$1t_h$
	BS	$3t_h$	$3t_h$	$3t_h$	$4t_h$	$5t_h$	$4t_h$	$6t_h$	$3t_h$	$200t_{enc} + 3t_h$	$3t_h$
	Sensor Cluster mode	–	–	–	–	–	–	–	–	–	–
Login	User (U_i)	–	$4t_h$	$4t_h + 2t_{kdf} + t_{mac} + t_{dec}$	$4t_h$	$5t_h$	$6t_h$	$7t_h$	$4t_h$	$4t_h + t_{enc}$	$6t_h + t_{enc}$
	BS	t_h	$4t_h$	$3t_h + 2t_{kdf} + t_{mac} + t_{enc}$	$6t_h$	$5t_h$	$5t_h$	$2t_h$	$5t_h$	$3t_h + t_{dec} + 2t_{enc}$	$5t_h + t_{dec} + t_{enc}$
Authentication	Sensor	$3t_h$	t_h	$2t_{kdf} + 2t_{mac} + t_{enc} + t_{dec}$	t_h	t_h	$2t_h$	t_h	t_h	–	–
	Cluster head	–	–	–	–	–	–	$8t_h$	–	$2t_h + t_{dec}$	$4t_h + 2t_{dec} + t_{enc}$

Table 22.3 Performance comparison between the proposed scheme and other schemes

	Wong et al. (2006)	Das (2009)	Nyang and Lee (2009)	Huang et al. (2010)	He et al. (2009)	Vaidya et al. (2010)	Fan et al. (2010)	Chen and Shih (2010)	Das et al. (2011)	Proposed
Supports change password	No	No	No	Yes	Yes	Yes	No	No	Yes	Yes
Supports mutual authentication	No	No	No	No	No	Yes	Yes	Yes	No	Yes
Resists denial of service attack	No	No	No	No	No	No	Yes	No	No	Yes
Resilient against node capture attack	No	No	No	Yes	No	No	No	No	Yes	Yes
Establish secret session key between user and sensor mode/cluster head	No	No	No	No	No	No	Yes	No	No	Yes
Supports dynamic node addition	No	No	No	No	No	No	No	No	Yes	Yes

22.5.6 *Resisting Server (Base Station) Masquerade Attack*

In our protocol, the adversary masquerading as BS must frame a valid login message $E_{\mathrm{MKCHi}}\{\mathrm{ID}_i$, CIDu, e_i, u, P, T_i, T_1, $T_2\}$ where $P = h(N_i||T_2||T_i)$. As discussed above, the adversary does not know the values of MKCHu, T_i, T_1, T_2, T_i. Without knowing the above mentioned values, it is not possible for the adversary to generate the valid login message to cluster head. Hence the proposed protocol is secure against server (BS) masquerade attack.

22.6 Performance Evaluation

In this section, we analyze the communication and computation cost required by our protocol and we compare the same with relevant protocols. In the below tables, the following notations are used: $t_{\mathrm{p}i}$: public-key computation; t_{pr}: private-key computation; t_h: hash computation; t_{enc}: symmetric-key encryption; t_{dec}: symmetric-key decryption; t_{kdf}: key derivation function computation and t_{mac}: message authentication code (MAC) (Tables 22.2, 22.3).

22.7 Conclusion

In this paper, we have reviewed authentication scheme proposed by Das et al. for WSN. Although their scheme can withstand privileged-insider attack, we have shown that their scheme is still vulnerable to major cryptographic attacks. Consequently, we have proposed an improved scheme to eliminate such problems. Compared with related schemes, the proposed scheme requires less computation cost and resists all major cryptographic attacks. As a result, the proposed scheme is able to provide greater security and will be practical in wireless sensor network deployment.

References

1. Das, A.K., Pranay, S., Santanu, C., Jamuna, K.S.: A dynamic password-based user authentication scheme for hierarchical wireless sensor networks. Elsevier J. Netw. Comput. Appl. **2012**(35), 1646–1656 (2012)
2. Wong, K., Zheng, Y., Cao, J., Wang, S.: A dynamic user authentication scheme for wireless sensor networks. In: Proceedings of IEEE International Conference on Sensor Networks, Ubiquitous, and Trustworthy Computing, IEEE Computer Society, pp. 244–51. (2006)
3. Das, M.L.: Two-factor user authentication in wireless sensor networks. IEEE Trans. Wireless Commun. **8**(3), 1086–1090 (2009)
4. Nyang, D.H., Lee, M.K.: Improvement of Das's two-factor authentication protocol in wireless sensor networks. In: Cryptology ePrint Archive. Report 2009/631; 2009

5. Huang, H.F., Chang, Y.F., Liu, C.H.: Enhancement of two-factor user authentication in wireless sensor networks. In: Sixth International Conference on Intelligent Information hID¡ng and Multimedia Signal Processing, pp. 27–30. (2010)
6. He, D., Gao, Y., Chan, S., Chen, C., Bu, J.: An enhanced two-factor user authentication scheme in wireless sensor networks. AdHoc Sens. Wireless Netw. **10**(4), 361–371 (2010)
7. Vaidya, B., Makrakis, D., Mouftah, H.T.: Improved two-factor user authentication in wireless sensor networks. In: Second International Workshop on Network Assurance And Security Services in Ubiquitous Environments, pp. 600–606. (2010)
8. Fan, R., Ping, L.D., Fu, J.Q., Pan, X.Z.: A secure and efficient user authentication protocol for two-tieres wireless sensor networks. In: Second Pacific-Asia Conference on Circuits, Communications and System (PACCS2010), pp. 425–428. (2010)
9. Chen, T.H., Shih, W.-K.: A robust mutual authentication protocol for wireless sensor networks. ETRI J. **32**(5), 704–712 (2010)
10. Watro, R., Kong, D., Cuti, S., Gardiner, C., Lynn, C., Kruus, P, Tiny, P.K.: Securing sensor networks with public key technology. In: Proceedings of the 2nd ACM Workshop On Security Of Adhoc and Sensor Networks, SASN2004, pp. 59–64. Washington, DC, USA, Oct 2004
11. Das, A.K., Sengupta, I.: An effective group-based key establishment scheme for large-scale wireless sensor networks using bivariate polynomials. In: Proceedings of 3rd IEEE International Conference on Communication Systems Software and Middleware (COMSWARE2008), pp. 9–16. (2008)
12. Dong, Q., Liu, D.: Using auxiliary sensors for pairwise key establishment in WSN. In: Proceedings of IFIP International Conferences on Networking (Networking2007), Lecture notes in computer science (LNCS), vol. 4479, pp. 251–262. (2007)

Khan, M.K., [illegible]: [illegible] two-factor user authentication in wireless sensor networks. Sensors 10(3), [illegible] (2010)

[illegible]: [illegible] two-factor user authentication [illegible]. In: Second International Conference on [illegible] (2011)

Wong, K.H.M., Zheng, Y., Cao, J., Wang, S.: A dynamic user authentication scheme for wireless sensor networks. In: IEEE International Conference on Sensor Networks, Ubiquitous, and Trustworthy Computing (SUTC'06), pp. 244–251 (2006)

Yeh, H.L., Chen, T.H., Liu, P.C., Kim, T.H., Wei, H.W.: A secured authentication protocol for wireless sensor networks using elliptic curves cryptography. Sensors 11(5), 4767–4779 (2011)

About the Book

Wireless communication and sensor networks would form the backbone to create pervasive and ubiquitous environments that would have profound influence on the society and thus are important to the society. The wireless communication technologies and wireless sensor networks would encompass a wide range of domains such as HW devices such as motes, sensors and associated instrumentation, actuators, transmitters, receivers, antennas, etc., sensor network aspects such as topologies, routing algorithms, integration of heterogeneous network elements and topologies, designing RF devices and systems for energy efficiency and reliability etc. These sensor networks would provide opportunity to continuously and in a distributed manner monitor the environment and generate the necessary warnings and actions. However most of the developments have been demonstrated only in controlled and laboratory environments. So we are yet to see those powerful, ubiquitous applications for the benefit of the society.

The conference and consequentially the proceedings would provide opportunity to the researchers to interact with other researchers and share their researches covering all the above areas. The proceedings of the conference thus covers the research work of different authors in the area of wireless sensor networks, wireless communications, devices, tools and techniques for WSN, and applications of wireless sensor networks.

This book is beneficial for those researchers who are working in the area of wireless sensor networks, wireless communication, and developing applications of Wireless sensor networks.

R. Maringanti et al. (eds.), *Proceedings of Ninth International Conference on Wireless Communication and Sensor Networks*, Lecture Notes in Electrical Engineering 299,
DOI: 10.1007/978-81-322-1823-4, © Springer India 2014

About the Book

About the Editors

Prof. R. Maringanti is a Professor in the Indian Institute of Information Technology Allahabad. He has published 2 books and more than 70 research papers. He has been the advisor to a number of engineering colleges and universities in India and abroad. His areas of interest include Artificial Intelligence, Intelligent Systems, Digital Design, Embedded Systems, Machine Vision, Computer Based Instrumentation and Control, Automation, Computer and Sensor Networks, Computer Based Instructional Systems, Cognitive Sciences, Modelling and Simulation.

Dr. M.D. Tiwari is the founder Director of the Indian Institute of Information Technology Allahabad. He has published more than 15 books and about 200 research papers. He facilitated the establishment of All India Council of Technical Education and he played significant role in the development of Science and Technology in India and University Grant Commission, New Delhi. He was Vice-chancellor of MJP Rohilkhand University Bareilly, UP and first Chairman of UP State Council of Higher Education, Lucknow.

Prof. Anish Arora is a Professor of Computer Science in the Ohio State University, Columbus, USA. He is a pioneer in the area of wireless sensor networks and has implemented the wireless sensor networks in real life situation. He has to his credit more than 60 papers. He has a number of international collaborations. His research interests include Wireless sensor networks; fault-tolerant, secure and timely computing; distributed systems and networks; embedded systems networking; component-based design; formal methods; concurrency semantics; dynamical systems.

R. Maringanti et al. (eds.), *Proceedings of Ninth International Conference on Wireless Communication and Sensor Networks*, Lecture Notes in Electrical Engineering 299,
DOI: 10.1007/978-81-322-1823-4, © Springer India 2014

Author Index

R. Maringanti et al. (eds.), *Proceedings of Ninth International Conference on Wireless Communication and Sensor Networks*, Lecture Notes in Electrical Engineering 299,
DOI: 10.1007/978-81-322-1823-4, © Springer India 2014